Information Circular 9515

Compendium of Structural Testing Data for 20-psi Coal Mine Seals

By R. Karl Zipf, Jr., Ph.D., P.E., Eric S. Weiss, Samuel P. Harteis, P.E., and Michael J. Sapko

DEPARTMENT OF HEALTH AND HUMAN SERVICES
Centers for Disease Control and Prevention
National Institute for Occupational Safety and Health
Pittsburgh Research Laboratory
Pittsburgh, PA

August 2009

> This document is in the public domain and may be freely copied or reprinted.

Disclaimer

Mention of any company or product does not constitute endorsement by the National Institute for Occupational Safety and Health (NIOSH). In addition, citations to Web sites external to NIOSH do not constitute NIOSH endorsement of the sponsoring organizations or their programs or products. Furthermore, NIOSH is not responsible for the content of these Web sites. All Web addresses referenced in this document were accessible as of the publication date.

Ordering Information

To receive documents or other information about occupational safety and health topics, contact NIOSH at

> Telephone: **1–800–CDC–INFO** (1–800–232–4636)
> TTY: 1–888–232–6348
> e-mail: cdcinfo@cdc.gov
>
> or visit the NIOSH Web site at **www.cdc.gov/niosh**.

For a monthly update on news at NIOSH, subscribe to NIOSH *eNews* by visiting **www.cdc.gov/niosh/eNews**.

DHHS (NIOSH) Publication No. 2009–151

August 2009

SAFER • HEALTHIER • PEOPLE™

CONTENTS

Page

Abstract 1
Introduction 2
Experimental mine and test procedures 5
    Lake Lynn experimental mine (LLEM) 5
    Explosion tests in the LLEM 6
    Hydrostatic chamber tests in the LLEM 7
Instrumentation and data collection 8
    Pressure waves from test explosions 8
    Pressure and displacement measurement locations 9
    Loading conditions for seal tests 10
    Boundary conditions for seal tests 11
    Response times, time constants, and frequency responses for sensors used in the LLEM 11
    Data acquisition system characteristics 16
    Quality of pressure-time and displacement-time measurements 17
    Adequacy of pressure-time and displacement-time measurements for structural analysis 20
    Comments on smoothing 21
General construction details for category 1 through 6 seals 21
    Category 1 seals: concrete or concretelike materials with internal steel reinforcement
        and anchorage to rock 21
    Category 2 seals: pumpable cementitious materials with no steel reinforcement and
        no hitching 27
    Category 3 seals: articulated structures – solid and hollow-core concrete blocks with
        or without hitching 30
    Category 4 seals: polymer and aggregate materials without hitching 38
    Category 5 seals: wood-crib-block seals with or without hitching 40
    Category 6 seals: articulated structures – lightweight blocks with or without hitching 45
Structural testing data for 20-psi coal mine seals 49
Summary and conclusions 57
References 58
Appendix.—Detailed summary of seal structure tests and test data 60

ILLUSTRATIONS

1. Plan view of the Lake Lynn experimental mine (LLEM) 5
2. Plan view of the LLEM showing the multiple-entry area and the seal and stopping
    locations 6
3. Schematic of the hydrostatic chamber 7
4. Schematic of pressure and displacement measurement points for typical explosion tests
    in the LLEM 9
5. Dimensionless response function for instantaneous unit step input 12
6. Calculated pressure response for instantaneous 100-psi step input 13
7. Calculated pressure response for instantaneous 300-psi step input 14
8. Calculated displacement response for instantaneous 6-in step input (10-ms response time) 15
9. Calculated displacement response for instantaneous 6-in step input (75-ms response time) 16

CONTENTS—Continued

Page

10. Typical P-t and D-t data for the entire duration of an explosion test in the LLEM 17
11. Expanded time-scale view of P-t data from two separate pressure transducers 18
12. Expanded time-scale view of P-t data from the same pressure transducer recorded with separate data acquisition systems operating at 1,500 and 5,000 samples per second 18
13. Expanded time-scale view of D-t data from the same displacement transducer recorded with separate data acquisition systems operating at 1,500 and 5,000 samples per second 19
14. Computed D-t responses for hypothetical structure using P-t curves in Figure 12 as input for structural analysis 20
15. Front-, plan-, and side-view drawings of category 1A structure: Insteel 3-D seal from Precision Mine Repair, Inc. 22
16. Detailed plan-view drawing of category 1A structure: Insteel 3-D seal from Precision Mine Repair, Inc. 23
17. Closeup of a rear Insteel panel with Stayform backing and the horizontal #3 steel reinforcement bars 23
18. Insteel 3-D seal under construction showing the rear Insteel panel with stayform backing, one plane of vertical #8 steel reinforcement bars and anchors, the horizontal #8 steel reinforcement bar anchors, and the horizontal #3 steel reinforcement bars 24
19. Insteel 3-D seal under construction with the addition of the front Insteel panel and the front plane of vertical #8 steel reinforcement bars 25
20. Front-, plan-, and side-view drawings of category 1B structure: Meshblock seal from R. G. Johnson Co., Inc. 26
21. Closeup of Meshblock seal under construction showing the Meshblocks and the vertical #8 steel reinforcement bars and anchors 27
22. Meshblock seal under construction showing the lower Meshblocks, the vertical #8 steel reinforcement bars and anchors, and placement of shotcrete 27
23. Front-, plan-, and side-view drawings of category 2 structure made from pumpable cementitious materials with no steel reinforcement and no hitching 28
24. Formwork for a typical pumpable seal showing the vertical posts, the horizontal boards, and the brattice liner (partially removed) 29
25. Inside the formwork of a typical pumpable seal showing the brattice liner and the cementitious filling material 29
26. Front-, plan-, and side-view drawings of category 3A structure: standard solid-concrete-block seal with simulated hitching 31
27. Solid-concrete-block seal under construction showing the center pilaster and the fully mortared joints on all sides 32
28. In background, top of a solid-concrete-block seal showing small cut blocks and mortar filling at the seal top 32
29. Completed solid-concrete-block seal without a center pilaster showing the angle iron hitch on the ribs and floor only 33
30. Front-, plan-, and side-view drawings of category 3B structure: solid-concrete-block seal with Packsetter bags 34

CONTENTS—Continued

Page

31. Tongue-and-groove, solid-concrete-block seal with center pilaster using pressurized Packsetter grout bags in lieu of hitching around the seal perimeter 35
32. Front-, plan-, and side-view drawings of category 3C structure: solid- or hollow-core concrete block ventilation stoppings 36
33. Dry-stacked concrete block stopping showing wedges used to fit the stopping to ribs and small cut blocks and a wood plank at the top 37
34. Dry-stacked concrete block stopping showing application of an approved sealant to the surface 37
35. Front-, plan-, and side-view drawings of category 4 structure: polymer and aggregate seal 38
36. Polymer and aggregate seal from MICON showing the rear dry-stacked, hollow-core concrete block wall, the partially completed front wall, a polymer coating on the inside surface of the form walls, and an approved sealant on the outside surface of the form wall 39
37. Polymer and aggregate seal from MICON showing the rear and front form wall, the polymer coating on the inner surfaces of the form walls, and the polyurethane foam and aggregate mixture filling the inner core of the seal 40
38. Front-, plan-, and side-view drawings of category 5A structure: wood-crib-block seal with plywood facing and simulated hitching 41
39. Closeup of a hitched wood-crib-block seal showing the layer of rock dust placed between each layer of wood crib blocks 41
40. Construction of a hitched wood-crib-block seal without plywood facing showing the wood crib blocks separated by a rock dust layer, the angle iron hitching around the ribs and floor only, and the final coating with an approved sealant 42
41. Front-, plan-, and side-view drawings of category 5B structure: wood-crib-block seal with Packsetter bags 43
42. Construction of a wood-crib-block seal with pressurized Packsetter grout bags in lieu of hitching showing the application of glue between all wood-crib-block surfaces 44
43. Construction of a wood-crib-block seal with pressurized Packsetter grout bags in lieu of hitching showing the glued wood crib blocks and the grout bags used along the roof and ribs 44
44. Front-, plan-, and side-view drawings of category 6A structure: lightweight block seal, 24 in thick with hitching 46
45. Construction of a 24-in-thick Omega block seal with hitching and a center pilaster 46
46. Construction of an Omega block seal showing the wood board and wedges used to secure the seal at the roof 47
47. Completed 24-in-thick Omega block seal showing the angle iron hitch along the ribs and floor and the outer surface coated with an approved sealant 47
48. Front-, plan-, and side-view drawings of category 6B structure: lightweight block seal, 40 in thick without hitching 48
49. Construction of a 40-in-thick Omega block seal without hitching and no center pilaster 49

CONTENTS—Continued

Page

A-1. Category 1A - structure #1 - test 1 - static, nonuniform loading; Insteel 3-D seal - shotcrete with reinforcement - LLEM test #419 80

A-2. Category 1A - structure #2 - test 1 - static, nonuniform loading; Insteel 3-D seal - shotcrete with reinforcement - LLEM test #419 80

A-3. Category 1A - structure #3 - test 1 - static, nonuniform loading; Insteel 3-D seal - shotcrete with reinforcement - LLEM test #419 81

A-4. Category 1A - structure #4 - test 1 - static, nonuniform loading; Insteel 3-D seal - shotcrete with reinforcement - LLEM test #420 81

A-5. Category 1A - structure #5 - test 1 - static, nonuniform loading; Insteel 3-D seal - shotcrete with reinforcement - LLEM test #420 82

A-6. Category 1A - structure #6 - test 1 - static, nonuniform loading; Insteel 3-D seal - shotcrete with reinforcement - LLEM test #420 82

A-7. Category 1A - structure #7 - test 1 - static, uniform loading; Insteel 3-D seal - shotcrete with reinforcement - PR-1 83

A-8. Category 1B - structure #1 - tests 1 to 4 - static, nonuniform loading; Meshblock seal - shotcrete with reinforcement - LLEM tests #347–350 83

A-9. Category 1B - structure #1 - tests 5 to 8 - static, nonuniform loading; Meshblock seal - shotcrete with reinforcement - LLEM tests #351–358 84

A-10. Category 1B - structure #1 - tests 9 to 12 - static, nonuniform loading; Meshblock seal - shotcrete with reinforcement - LLEM tests #359–362 84

A-11. Category 1B - structure #1 - tests 13 to 16 - static, nonuniform loading; Meshblock seal - shotcrete with reinforcement - LLEM tests #363–366 85

A-12. Category 1B - structure #2 - tests 1 to 3 - static, nonuniform loading; Meshblock seal - shotcrete with reinforcement - LLEM tests #347–349 85

A-13. Category 1B - structure #2 - tests 4 and 5 - static, nonuniform loading; Meshblock seal - shotcrete with reinforcement - LLEM tests #350–351 86

A-14. Category 1B - structure #3 - tests 1 to 3 - static, nonuniform loading; Meshblock seal - shotcrete with reinforcement - LLEM tests #347–349 86

A-15. Category 1B - structure #4 - tests 1 and 2 - static, nonuniform loading; Meshblock seal - shotcrete with reinforcement - LLEM tests #347–348 87

A-16. Category 1B - structure #5 - tests 1 and 2. Meshblock seal - shotcrete with reinforcement - LLEM tests #347–348 87

A-17. Category 1B - structure #6 - tests 1 and 2 - static, nonuniform loading; Meshblock seal - shotcrete with reinforcement - LLEM tests #350–351 88

A-18. Category 2A - structure #1 - test 1 - reflected, uniform loading; pumpable 48 in - LLEM test #508 88

A-19. Category 2A - structure #1 - test 2 - reflected, uniform loading; pumpable 48 in - LLEM test #509 89

A-20. Category 2A - structure #2 - test 1 - static, uniform loading; pumpable 48 in - test C3-44E 89

A-21. Category 2A - structure #3 - test 1 - static, uniform loading; pumpable 48 in - test C7-64W 90

CONTENTS—Continued

Page

A-22. Category 2A - structure #3 - test 2 - static, uniform loading; pumpable 48 in - test C7-68W 90
A-23. Category 2A - structure #3 - test 3 - static, uniform loading; pumpable 48 in - test C7-70W 91
A-24. Category 2A - structure # 4 - test 1 - static, uniform loading; pumpable 48 in - test L2-51E 91
A-25. Category 2B - structure #1 - tests 1 and 2 - static, nonuniform loading; pumpable 36 in - LLEM tests #354–355 92
A-26. Category 2C - structure #1 - tests 1 and 2 - static, nonuniform loading; pumpable 24 in - LLEM tests #354–355 92
A-27. Category 2C - structure #2 - tests 1 and 2 - static, nonuniform loading; pumpable 24 in - LLEM tests #354–355 93
A-28. Category 2C - structure #3 - tests 1 and 2 - static, nonuniform loading; pumpable 24 in - LLEM tests #354–355 93
A-29. Category 2C - structure #4 - test 1 - static, nonuniform loading; pumpable 24 in - LLEM test #403 94
A-30. Category 2C - structure #4 - test 2 - static, nonuniform loading; pumpable 24 in - LLEM test #404 94
A-31. Category 2C - structure #4 - test 3 - static, nonuniform loading; pumpable 24 in - LLEM test #405 95
A-32. Category 2C - structure #4 - test 4 - static, nonuniform loading; pumpable 24 in - LLEM test #406 95
A-33. Category 3A - structure #1 - test 1 - static, nonuniform loading; standard solid-concrete-block seal - LLEM test #403 96
A-34. Category 3A - structure #1 - test 2 - static, nonuniform loading; standard solid-concrete-block seal - LLEM test #404 96
A-35. Category 3A - structure #1 - test 3 - static, nonuniform loading; standard solid-concrete-block seal - LLEM test #405 97
A-36. Category 3A - structure #1 - test 4 - static, nonuniform loading; standard solid-concrete-block seal - LLEM test #406 97
A-37. Category 3A - structure #2 - tests 1 to 6 - static, nonuniform loading; standard solid-concrete-block seal - LLEM tests #500–505 98
A-38. Category 3A - structure #2 - tests 7 to 10 - static, nonuniform loading; standard solid-concrete-block seal - LLEM tests #506–509 98
A-39. Category 3A - structure #3 - test 1 - static, nonuniform loading; standard solid-concrete-block seal - LLEM test #506 99
A-40. Category 3A - structure #3 - test 2 - static, nonuniform loading; standard solid-concrete-block seal - LLEM test #507 99
A-41. Category 3A - structure #4 - test 1 - static, uniform loading; standard solid-concrete-block seal - test C1-5E 100
A-42. Category 3A - structure #4 - test 2 - static, uniform loading; standard solid-concrete-block seal - test C1-8E 100

CONTENTS—Continued

Page

A-43. Category 3A - structure #4 - test 3 - static, uniform loading; standard solid-concrete-block seal - test C1-9E 101
A-44. Category 3A - structure #4 - test 4 - static, uniform loading; standard solid-concrete-block seal - test C1-10E 101
A-45. Category 3A - structure #4 - test 5 - static, uniform loading; standard solid-concrete-block seal - test C1-11E 102
A-46. Category 3A - structure #5 - test 1 - static, uniform loading; standard solid-concrete-block seal - test C6-60W 102
A-47. Category 3A - structure #5 - test 2 - static, uniform loading; standard solid-concrete-block seal - test C6-62E 103
A-48. Category 3A - structure #6 - test 1 - static, uniform loading; standard solid-concrete-block seal - test L1-37E 103
A-49. Category 3A - structure #7 - test 1 - static, uniform loading; standard solid-concrete-block seal - test SRCM 1 104
A-50. Category 3B - structure #1 - test 1 - static, nonuniform loading; solid-concrete-block seal with Packsetter bags - LLEM test #365 104
A-51. Category 3B - structure #2 - tests 1 and 2 - static, nonuniform loading; solid-concrete-block seal with Packsetter bags - LLEM tests #365–366 105
A-52. Category 3B - structure #3 - tests 1 and 2 - static, nonuniform loading; solid-concrete-block seal with Packsetter bags - LLEM tests #365–366 105
A-53. Category 3C - structure #1 - test 1 - static, nonuniform loading; hollow-core concrete-block ventilation stopping - LLEM test #427 106
A-54. Category 3C - structure #1 - test 2 - static, nonuniform loading; hollow-core concrete-block ventilation stopping - LLEM test #428 106
A-55. Category 3C - structure #2 - test 1 - static, nonuniform loading; hollow-core concrete-block ventilation stopping - LLEM test #427 107
A-56. Category 3C - structure #2 - test 2 - static, nonuniform loading; hollow-core concrete-block ventilation stopping - LLEM test #428 107
A-57. Category 3C - structure #3 - test 1 - static, nonuniform loading; hollow-core concrete-block ventilation stopping - LLEM test #427 108
A-58. Category 3C - structure #3 - test 2 - static, nonuniform loading; hollow-core concrete-block ventilation stopping - LLEM test #428 108
A-59. Category 3C - structure #3 - test 3 - static, nonuniform loading; hollow-core concrete-block ventilation stopping - LLEM test #429 109
A-60. Category 3C - structure #3 - test 4 - static, nonuniform loading; hollow-core concrete-block ventilation stopping - LLEM test #430 109
A-61. Category 3C - structure #3 - test 5 - static, nonuniform loading; hollow-core concrete-block ventilation stopping - LLEM test #432 110
A-62. Category 3C - structure #3 - test 6 - static, nonuniform loading; hollow-core concrete-block ventilation stopping - LLEM test #433 110
A-63. Category 3C - structure #4 - test 1 - static, nonuniform load; hollow-core concrete-block ventilation stopping - LLEM test #427 111

CONTENTS—Continued

Page

A-64. Category 3C - structure #4 - test 2 - static, nonuniform load; hollow-core concrete-block ventilation stopping - LLEM test #428 111

A-65. Category 3C - structure #4 - test 3 - static, nonuniform load; hollow-core concrete-block ventilation stopping - LLEM test #429 112

A-66. Category 3C - structure #4 - test 4 - static, nonuniform load; hollow-core concrete-block ventilation stopping - LLEM test #430 112

A-67. Category 3C - structure #4 - test 5 - static, nonuniform load; hollow-core concrete-block ventilation stopping - LLEM test #432 113

A-68. Category 3C - structure #4 - test 6 - static, nonuniform load; hollow-core concrete-block ventilation stopping - LLEM test #433 113

A-69. Category 3C - structure #4 - test 7 - static, nonuniform load; hollow-core concrete-block ventilation stopping - LLEM test #434 114

A-70. Category 3C - structure #5 - test 1 - static, nonuniform loading; solid-concrete-block ventilation stopping - LLEM test #457 114

A-71. Category 3C - structure #5 - test 2 - static, nonuniform loading; solid-concrete-block ventilation stopping - LLEM test #458 115

A-72. Category 3C - structure #5 - test 3 - static, nonuniform loading; solid-concrete-block ventilation stopping - LLEM test #459 115

A-73. Category 3C - structure #5 - test 4 - static, nonuniform loading; solid-concrete-block ventilation stopping - LLEM test #460 116

A-74. Category 3C - structure #5 - test 5 - static, nonuniform loading; solid-concrete-block ventilation stopping - LLEM test #461 116

A-75. Category 3C - structure #5 - test 6 - static, nonuniform loading; solid-concrete-block ventilation stopping - LLEM test #462 117

A-76. Category 3C - structure #6 - test 1 - static, nonuniform loading; solid-concrete-block ventilation stopping - LLEM test #457 117

A-77. Category 3C - structure #6 - test 2 - static, nonuniform loading; solid-concrete-block ventilation stopping - LLEM test #458 118

A-78. Category 3C - structure #6 - test 3 - static, nonuniform loading; solid-concrete-block ventilation stopping - LLEM test #459 118

A-79. Category 3C - structure #6 - test 4 - static, nonuniform loading; solid-concrete-block ventilation stopping - LLEM test #460 119

A-80. Category 3C - structure #6 - test 5 - static, nonuniform loading; solid-concrete-block ventilation stopping - LLEM test #461 119

A-81. Category 3C - structure #6 - test 6 - static, nonuniform loading; solid-concrete-block ventilation stopping - LLEM test #462 120

A-82. Category 3C - structure #6 - test 7 - static, nonuniform loading; solid-concrete-block ventilation stopping - LLEM test #463 120

A-83. Category 3C - structure #7 - tests 1 to 3 - static, nonuniform loading; solid-concrete-block ventilation stopping - LLEM tests #510–512 121

A-84. Category 3C - structure #7 - tests 4 to 6 - static, nonuniform loading; solid-concrete-block ventilation stopping - LLEM tests #513–515 121

CONTENTS—Continued

Page

A-85. Category 3C - structure #7 - tests 7 to 10 - static, nonuniform loading; solid-concrete-block ventilation stopping - LLEM tests #516–519 122

A-86. Category 3C - structure #8 - test 1 - static, nonuniform loading; solid-concrete-block ventilation stopping - LLEM test #510 122

A-87. Category 3C - structure #8 - test 2 - static, nonuniform loading; solid-concrete-block ventilation stopping - LLEM test #511 123

A-88. Category 3C - structure #8 - test 3 - static, nonuniform loading; solid-concrete-block ventilation stopping - LLEM test #512 123

A-89. Category 3C - structure #8 - test 4 - static, nonuniform loading; solid-concrete-block ventilation stopping - LLEM test #513 124

A-90. Category 3C - structure #8 - test 5 - static, nonuniform loading; solid-concrete-block ventilation stopping - LLEM test #514 124

A-91. Category 3C - structure #8 - test 6 - static, nonuniform loading; solid-concrete-block ventilation stopping - LLEM test #515 125

A-92. Category 3C - structure #8 - test 7 - static, nonuniform loading; solid-concrete-block ventilation stopping - LLEM test #516 125

A-93. Category 3C - structure #8 - test 8 - static, nonuniform loading; solid-concrete-block ventilation stopping - LLEM test #517 126

A-94. Category 3C - structure #8 - test 9 - static, nonuniform loading; solid-concrete-block ventilation stopping - LLEM test #518 126

A-95. Category 3C - structure #8 - test 10 - static, nonuniform loading; solid-concrete-block ventilation stopping - LLEM test #519 127

A-96. Category 4 - structure #1 - test 1 - static, uniform loading; polymer and aggregate seal - test C8 127

A-97. Category 4 - structure #1 - test 1 - static, uniform loading; polymer and aggregate seal - test C8 128

A-98. Category 4 - structure #1 - test 1 - static, uniform loading; polymer and aggregate seal - test C8 128

A-99. Category 4 - structure #1 - test 1 - static, uniform loading; polymer and aggregate seal - test C8 129

A-100. Category 5B - structure #1 - test 1 - static, nonuniform loading; wood-crib-block seal with Packsetter bags - LLEM test #396 129

A-101. Category 5B - structure #1 - test 2 - static, nonuniform loading; wood-crib-block seal with Packsetter bags - LLEM test #399 130

A-102. Category 6A - structure #3 - test 1 - static, nonuniform loading; lightweight blocks - 24 in with hitching - LLEM test #508 130

A-103. Category 6A - structure #3 - test 2 - static, nonuniform loading; lightweight blocks - 24 in with hitching - LLEM test #509 131

A-104. Category 6A - structure #2 - test 1 - static, uniform loading; lightweight blocks - 24 in with hitching - test 4-48 131

A-105. Category 6B - structure #1 - test 1 - static, nonuniform loading; lightweight blocks - 40 in, no hitching - LLEM test #403 132

CONTENTS—Continued

Page

A-106. Category 6B - structure #1 - test 2 - static, nonuniform loading; lightweight blocks - 40 in, no hitching - LLEM test #404 132

A-107. Category 6B - structure #1 - test 3 - static, nonuniform loading; lightweight blocks - 40 in, no hitching - LLEM test #405 133

A-108. Category 6B - structure #1 - test 4 - static, nonuniform loading; lightweight blocks - 40 in, no hitching - LLEM test #406 133

A-109. Category 6B - structure #2 - test 1 - static, nonuniform loading; lightweight blocks - 40 in, no hitching - LLEM test #501 134

A-110. Category 6B - structure #2 - test 2 - static, nonuniform loading; lightweight blocks - 40 in, no hitching - LLEM test #502 134

A-111. Category 6B - structure #2 - test 3 - static, nonuniform loading; lightweight blocks - 40 in, no hitching - LLEM test #503 135

A-112. Category 6B - structure #2 - test 4 - static, nonuniform loading; lightweight blocks - 40 in, no hitching - LLEM test #504 135

A-113. Category 6B - structure #2 - test 5 - static, nonuniform loading; lightweight blocks - 40 in, no hitching - LLEM test #505 136

A-114. Category 6B - structure #2 - test 6 - static, nonuniform loading; lightweight blocks - 40 in, no hitching - LLEM test #506 136

A-115. Category 6B - structure #2 - test 7 - static, nonuniform loading; lightweight blocks - 40 in, no hitching - LLEM test #507 137

A-116. Category 6B - structure #2 - tests 8 and 9 - static, nonuniform loading; lightweight blocks - 40 in, no hitching - LLEM tests #508–509 137

A-117. Category 6B - structure #3 - test 1 - static, nonuniform loading; lightweight blocks - 40 in, no hitching - LLEM test #501 138

A-118. Category 6B - structure #3 - test 2 - static, nonuniform loading; lightweight blocks - 40 in, no hitching - LLEM test #502 138

A-119. Category 6B - structure #4 - test 1 - static, nonuniform loading; lightweight blocks - 40 in, no hitching - LLEM test #503 139

A-120. Category 6B - structure #4 - test 2 - static, nonuniform loading; lightweight blocks - 40 in, no hitching - LLEM test #504 139

A-121. Category 6B - structure #4 - test 3 - static, nonuniform loading; lightweight blocks - 40 in, no hitching - LLEM test #505 140

A-122. Category 6B - structure #5 - test 1 - reflected, uniform loading; lightweight blocks - 40 in, no hitching - LLEM test #502 140

A-123. Category 6B - structure #6 - test 1 - reflected, uniform loading; lightweight blocks - 40 in, no hitching - LLEM test #503 141

A-124. Category 6B - structure #6 - test 2 - reflected, uniform loading; lightweight blocks - 40 in, no hitching - LLEM test #504 141

A-125. Category 6B - structure #6 - test 3 - reflected, uniform loading; lightweight blocks - 40 in, no hitching - LLEM test #505 142

A-126. Category 6B - structure #7 - test 1 - reflected, uniform loading; lightweight blocks - 40 in, no hitching - LLEM test #506 142

A-127. Category 6B - structure #8 - test 1 - static, uniform loading; lightweight blocks - 40 in, no hitching - test C5-53E 143

CONTENTS—Continued

TABLES

Page

1. Summary of seal types and structural testing data for 20-psi seal designs. 3
2. Total number and distribution of different seal categories as of November 2006 4
3. Response time, time constant, and frequency response for various pressure transducers used to record P-t data during LLEM tests 13
4. Response time, time constant, and frequency response for displacement transducers used to record D-t data during LLEM tests 15
5. Summary of category 1 seal structures: concrete with steel reinforcement 51
6. Summary of category 2 seal structures: pumpable cementitious materials 52
7. Summary of category 3 seal structures: articulated structures 54
8. Summary of category 4 seal structures: polymer and aggregate structures 54
9. Summary of category 5 seal structures: wood-crib-block structures 55
10. Summary of category 6 seal structures: lightweight block structures 56
A-1. Detailed summary of seal structure tests 60

ACRONYMS AND ABBREVIATIONS USED IN THIS REPORT

ASTM	American Society for Testing and Materials
CFR	Code of Federal Regulations
D-t	displacement-time
KS	Kinetic Systems
LLEM	Lake Lynn Experimental Mine
LVDT	linear variable displacement transducer
MSHA	Mine Safety and Health Administration
NI	National Instruments Corp.
NIOSH	National Institute for Occupational Safety and Health
P-t	pressure-time
RMR	Rock Mass Rating
SRCM	Safety Research Coal Mine
WAC	Wall Analysis Code
X	crosscut (e.g., "X-1" stands for "crosscut 1")

UNIT OF MEASURE ABBREVIATIONS USED IN THIS REPORT

ft	foot
ft^2	square foot
ft/s	foot per second
gal	gallon
hr	hour
Hz	hertz
in	inch
lb	pound
lb/ft^3	pound per cubic foot
m	meter
min	minute
ms	millisecond
pcf	pound per cubic foot
psi	pound-force per square inch
psig	pound-force per square inch gauge
sec	second
t	ton

COMPENDIUM OF STRUCTURAL TESTING DATA FOR 20-psi COAL MINE SEALS

By R. Karl Zipf, Jr., Ph.D., P.E.,[1] Eric S. Weiss,[2] Samuel P. Harteis, P.E.,[3] and Michael J. Sapko[4]

ABSTRACT

This report presents nearly all structural data available from explosion tests of 20-psi mine ventilation seals and concrete-block ventilation stoppings that were conducted by the National Institute for Occupational Safety and Health during 1997–2008. Although the seals tested were designed to meet the former federal 20-psi pressure design standard, the structural information contained herein on these seal tests will facilitate the analysis and design of coal mine seals that meet the new explosion pressure design criteria of 50 and 120 psi as set forth in the Mine Safety and Health Administration (MSHA)'s final rule on "Sealing of Abandoned Areas."

The seal testing data are organized into six broad categories of seal structures based on the materials used and the construction method for those 20-psi seals:

1. Concretelike materials with steel reinforcement and reinforcement bar anchorage to rock
2. Pumpable cementitious materials of varying compressive strengths with no steel reinforcement and no hitching
3. Articulated structures such as solid-concrete-block seals and ventilation stoppings made of solid and hollow-core concrete blocks
4. Polymer and aggregate materials without hitching
5. Wood-crib-block seals with or without hitching
6. Articulated structures such as lightweight blocks with or without hitching

This summary contains data on 52 different structures in the above categories—44 seals and 8 ventilation stoppings. The structural data sets include the applied loading on the tested seal represented by a pressure-time curve and, when available, the measured seal response represented by a displacement-time curve. The structural data sets enable the calibration and verification of numerical models of seal behavior at the 20-psi level, which may then facilitate future structural analyses of seal designs for the new 50- and 120-psi explosion pressure design criteria.

---

[1]Senior Research Mining Engineer, Pittsburgh Research Laboratory, National Institute for Occupational Safety and Health, Pittsburgh, PA.
[2]Team Leader (Senior Research Mining Engineer), Lake Lynn Laboratory Section, Pittsburgh Research Laboratory, National Institute for Occupational Safety and Health, Pittsburgh, PA.
[3]Research Mining Engineer, Pittsburgh Research Laboratory, National Institute for Occupational Safety and Health, Pittsburgh, PA.
[4]Principal Research Physical Scientist (retired), Pittsburgh Research Laboratory, National Institute for Occupational Safety and Health, Pittsburgh, PA.

INTRODUCTION

Seals are barriers constructed in underground coal mines throughout the United States to isolate abandoned mining areas from the active workings. Prior to the Sago Mine disaster in 2006, federal regulations required seals to withstand a 20-psi explosion pressure. On April 18, 2008, the Mine Safety and Health Administration (MSHA) issued "Sealing of Abandoned Areas; Final Rule," which includes requirements for seal strength, design, and construction of seals [73 Fed. Reg.[5] 21182 (2008)]. In the final rule [30 CFR[6] 75.335(a)], seals must:

(1) Withstand 50 psi if the sealed area is monitored and maintained inert;
(2) Withstand 120 psi if the sealed area is not monitored; or
(3) Withstand greater than 120 psi if the area is not monitored and certain conditions exist that might lead to higher explosion pressure.

30 CFR 75.335(b)(1) specifies the content of an engineering design application for seals. The design application must address the pressure-time (P-t) curve, engineering design and analysis, material properties, and other pertinent factors. To facilitate the analysis and design of seal structures that meet the new explosion pressure criterion, this report presents all structural data available from explosion tests conducted by NIOSH during 1997–2008 on seals designed to meet the former 20-psi pressure design standard.

This report organizes and presents the applied loading or P-t curves and, when available, the measured displacement-time (D-t) curves for 44 different seal structures tested prior to 2006 when the former 20-psi explosion pressure design criterion applied to mine seals. Also included in this data set are the applied loading P-t curves and response D-t curves for eight different ventilation stoppings constructed with solid or hollow-core concrete blocks. These structural test results against the stoppings are included as supplemental information pertinent to the design of seals that incorporate concrete blocks in some capacity.

Table 1 summarizes the testing program for seals designed to meet the former 20-psi pressure design standard for seals as conducted by the National Institute for Occupational Safety and Health (NIOSH) during 1997–2008. The program included tests on six broad categories of seal structures organized by the main seal construction material used and the construction method.

Category 1 includes seals made of concrete or concretelike materials such as shotcrete or gunite with internal steel reinforcement and anchorage to surrounding rock via additional steel reinforcement bars. Seals in this category are the Insteel 3-D seal (Precision Mine Repair, Inc., Ridgway, IL) and the Meshblock seal (Tecrete Industries Pty. Ltd., New South Wales, Australia, and R. G. Johnson Co., Inc., Washington, PA).

Category 2 includes the so-called pumpable seals constructed with different thicknesses of pumpable cementitious material depending on its compressive strength. Category 2 seals do not contain internal steel reinforcement and are not hitched[7] to the surrounding rock except through friction between the seal material and the rock. Manufacturers of seals in this category include Minova (Georgetown, KY), HeiTech Corp. (Cedar Bluff, VA, and Morgantown, WV), and R. G. Johnson Co., Inc. (Washington, PA).

---

[5] *Federal Register.* See Fed. Reg. in references.
[6] *Code of Federal Regulations.* See CFR in references.
[7] "Hitching" a seal involves constructing a foundation for the seal, usually by excavation or trenching into competent floor rock and rib coal.

Table 1.—Summary of seal types and structural testing data for 20-psi seal designs

Structure type	Total No. of structures tested	No. of structures tested with multiple loads	No. of structures tested to failure	No. of tests with P-t data only	No. of tests with both P-t and D-t data
CATEGORY 1: Concrete or concretelike materials with internal steel reinforcement and anchorage to rock					
1A. Insteel 3-D seal	7	0	0	0	7
1B. Meshblock seal	6	6	5	30	0
CATEGORY 2: Pumpable cementitious materials with no steel reinforcement and no hitching					
2A. Compressive strength: 200 psi, >48 in thick	4	2	3	0	7
2B. Compressive strength: 433 psi, >36 in thick	1	1	0	2	0
2C. Compressive strength: 480–600 psi, 24–30 in thick	4	4	0	6	4
CATEGORY 3: Articulated structures: solid and hollow-core concrete blocks with or without hitching					
3A. Standard solid-concrete-block seal with hitching	7	5	3	11	14
3B. Solid-concrete-block seal with Packsetter Bags and without hitching	3	3	1	5	0
3C. Ventilation stoppings: solid and hollow-core concrete blocks	8	8	7	10	40
CATEGORY 4: Polymer and aggregate materials without hitching					
4. Polymer and aggregate materials	1	0	1	0	1 (11 LVDTs)
CATEGORY 5: Wood-crib-block seals with or without hitching					
5A. Wood-crib-block seal with hitching	0	0	0	0	0
5B. Wood-crib-block seal with glue and Packsetter Bags	1	1	0	0	2
CATEGORY 6: Articulated structures: lightweight blocks with or without hitching					
6A. Lightweight blocks: 24 in thick with hitching	2	1	2	0	3
6B. Lightweight blocks: 40 in thick without hitching	8	5	6	2	22
TOTAL	52	36	28	66	100

Category 3 seals are "articulated" structures made of discrete concrete blocks, either solid or hollow-core, which may or may not be hitched to the surrounding rock. Seals in this category are the standard solid-concrete-block seal, which required hitching, and the solid-concrete-block seal with Packsetter Bags supplied by Strata Mine Services (Strata Products Worldwide, LLC, Marietta, GA), which does not require hitching to withstand 20 psi. Also included in this category are ventilation stoppings designed to withstand 2-psi overpressure, which are made of solid or hollow-core concrete blocks and do not require hitching.

Category 4 seals are made from polymer mixed with dry, crushed limestone aggregate, ranging in size from 0.25 to 1 in, placed between two, dry-stacked, hollow-core or solid-concrete-block form walls. This seal does not require hitching. The only example is the MICON 550 seal (MICON, Glassport, PA).

Category 5 seals are made from stacked wood crib blocks nailed together with 4-in-long nails. These seals, used where high convergence is expected, require hitching into the surrounding rock. This category also includes wood-crib-block seals that were glued together. The use of Packsetter Bags supplied by Strata Mine Services eliminated the requirement for hitching.

Category 6 seals are made from lightweight Omega blocks (Burrell Mining Products International, Inc., New Kensington, PA), cemented together with an MSHA-approved bonding agent called BlocBond, product No. 1225-51, a fiber-reinforced surface bonding cement manufactured by Quikrete Co., Atlanta, GA. Lightweight block seals constructed 24 or 32 in thick required hitching, whereas lightweight block seals more than 40 in thick did not require hitching. Seals constructed from lightweight blocks are no longer permitted, but the data are included herein for completeness.

This summary of NIOSH seal tests contains data on a total of 52 different structures including 44 seals and 8 stoppings. As shown in Table 1, many of the structures were subject to multiple loadings, and many of the structures were tested to failure. In some cases, the applied explosion pressure severely damaged the structure or collapsed it completely in the first test. In most cases (36 of the 52), the structures were subjected to multiple explosion loads. Twenty-eight of the fifty-two structures tested were loaded to failure. Finally, as indicated in Table 1, most of the structures considered herein have measured response data in the form of a D-t curve from a linear variable displacement transducer (LVDT).

Of the 52 different structures tested (44 seals and 8 stoppings), 41 were tested via explosion tests conducted at NIOSH's Lake Lynn Experimental Mine (LLEM). In addition to the explosion tests, 11 different structures were tested in one of the hydrostatic chambers located in the LLEM—9 in the small hydrostatic test chamber and 2 in the large chamber. Of the 15 tests reported with the small hydrostatic chamber in the LLEM, only 5 used water pressure as the loading medium. The other 10 tests used a confined methane-air or similar gaseous mixture explosion within the hydrostatic chamber to develop the test pressure. Both of the tests in the large hydrostatic test chamber used a confined gas explosion to develop the test pressure. One test was conducted in an experimental chamber in the Safety Research Coal Mine (SRCM) at the NIOSH Pittsburgh Research Laboratory. This test used water pressure to develop the applied loading.

NIOSH researchers also obtained data compiled by MSHA on the approximate number and seal type of all of the 20-psi seals in existence as of November 2006. Table 2 presents those numbers and the percentage of seals in each category.

Table 2.—Total number and distribution of different seal categories as of November 2006

Seal type/ category	Description	Number	Percent of total
1	Concrete or concretelike materials with internal steel reinforcement and anchorage to rock	2,602	20
2	Pumpable cementitious material with no steel reinforcement and no hitching	3,153	24
3	Articulated structures: solid and hollow-core concrete blocks with or without hitching	2,692	21
4	Polymer and aggregate materials without hitching	1,208	9
5	Wood-crib-block seals with or without hitching	190	1
6	Articulated structures: lightweight blocks with or without hitching	3,210	25
TOTAL		13,055	100

EXPERIMENTAL MINE AND TEST PROCEDURES

Lake Lynn Experimental Mine (LLEM)

The structural tests on coal mine seals and stoppings were conducted at the NIOSH Lake Lynn Laboratory [Mattes et al. 1983; Triebsch and Sapko 1990]. Lake Lynn is one of the world's foremost mining laboratories for conducting large-scale surface and underground research in mining health and safety technology. It is located about 50 miles southeast of Pittsburgh, near Fairchance, Fayette County, PA, and occupies a former limestone mine.

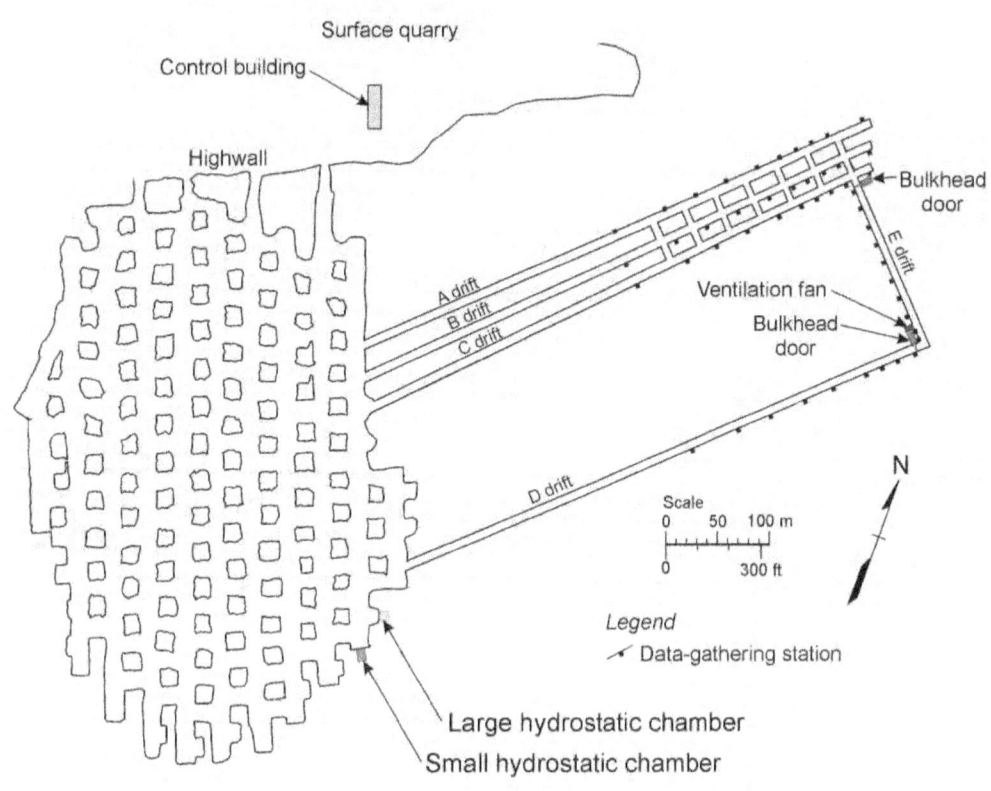

Figure 1.—Plan view of the Lake Lynn Experimental Mine (LLEM).

The underground LLEM (Figure 1) is unique in that it can simulate current U.S. coal mine geometries for a variety of mining scenarios, including multiple-entry room-and-pillar mining and longwall mining. The old limestone mine workings are shown on the left in Figure 1. Five new drifts (horizontal passageways in a mine) were developed to simulate the geometries of typical U.S. coal mines. The LLEM has four parallel drifts: A, B, C, and D. Drifts C and D are connected by E-drift, a 500-ft-long drift that simulates a longwall face. D-drift is a 1,640-ft-long single entry that can be separated from E-drift by an explosion-resistant bulkhead door. Drifts A, B, and C simulate longwall gate road entries or room-and-pillar workings. These three drifts are each approximately 1,600 ft long, with seven crosscuts at the inby end. A second explosion-resistant bulkhead door is used to separate the multiple entries from E-drift at the intersection with C-drift.

Explosion tests can be conducted in the single-entry D-drift; the multiple-entry area of A-, B-, and C-drifts; or various other configurations including the longwall E-drift. The entries are about 20 ft wide by about 6.5 ft high with cross-sectional areas of 130–140 ft^2. The crosscuts are 17–19 ft wide and about 7.2 ft high with a cross-sectional area of about 130 ft^2.

From August 1983 (when the first explosion test was conducted) to July 2008, a total of 527 consecutively numbered explosion tests were conducted in the LLEM.

Explosion Tests in the LLEM

Figure 2 shows an expanded view of the test area in the multiple-entry section of the LLEM. The faces, or inby (closed) ends, of A-, B-, and C-drifts are on the right in the figure. For most of the seal and stopping tests, the explosions were conducted in C-drift and the structures were built in crosscuts 1, 2, and 3 between B- and C-drifts, as shown in Figure 2. The evaluation of one type of seal was conducted in A-drift with the seals located in the crosscuts between A- and B-drifts. The evaluations of some of the ventilation stoppings were also conducted in A-drift as part of another explosion program with the stoppings located in X-6[8] and X-7 between A- and B-drifts and seals located in X-1 through X-5. For clarity, the A-drift testing scenarios are not shown in Figure 2.

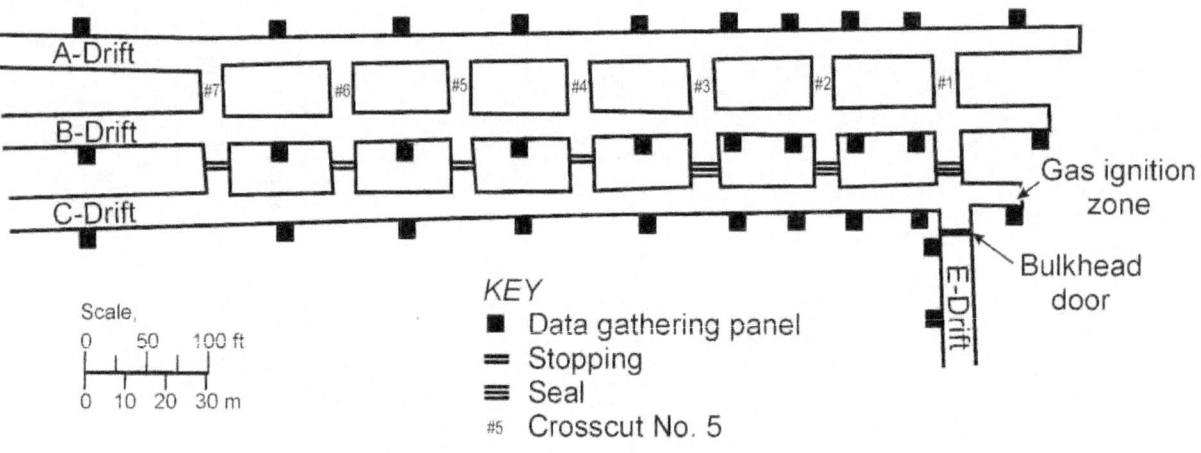

Figure 2.—Plan view of the LLEM showing the multiple-entry area and the seal and stopping locations. The first crosscut, designated as "#1", is nearest the dead end of drifts A, B, or C.

Before each explosion test, a 60-t pneumatically operated, track-mounted, concrete and steel bulkhead was positioned near the intersection of C- and E-drifts to contain the explosion pressures within the multiple-entry area. The LLEM bulkhead door and some of the other infrastructure were designed to withstand explosion overpressures of up to 100 psi. Higher pressures have been recorded at areas away from these structures.

---

[8]The abbreviation "X" stands for "crosscut" throughout this report, e.g., "X-1" stands for "crosscut 1."

For the LLEM explosion tests, natural gas was injected into the ignition zone. This natural gas is composed of ~97%–98% methane, ~1.5% ethane, and small percentages of other higher-order hydrocarbons. Sample lines within the ignition zone were used to draw gas samples to an infrared analyzer on the surface for measurement of the methane concentration. In addition, samples were collected in evacuated test tubes and sent to verify the analyses using gas chromatography. Most of the tests used a ~9%–10% methane-air concentration within an ignition zone contained in the C-drift face area with a clear plastic diaphragm. A few of the tests used a larger gas ignition zone. A fan with an explosion-proof motor housing mixed the natural gas and air prior to ignition. Electrically activated matches located either at the face (closed end) or outby the face within the gas ignition zone, depending on the explosion overpressure desired, were used to ignite the methane-air mixtures. In some of the tests, shelves of pulverized bituminous coal dust were also suspended in the drifts as a means to increase the explosion overpressures. For each of these explosion tests, the gas was ignited and the explosion pressure traveled out C-drift. For the explosion tests conducted in A-drift, the length of the gas ignition zone was varied to obtain higher total explosion overpressures at the stopping locations, i.e., the methane-air concentration was contained within a 50- or 85-ft-long gas ignition zone (as measured from the closed end of A-drift) for the different tests.

Hydrostatic Chamber Tests in the LLEM

Two hydrostatic chambers located within the high-roof section of the LLEM beyond the mouth of D-drift (Figure 1) enable researchers to impart pneumatic, hydrostatic, or explosion pressure loadings on test seals. Figure 3 is a schematic of the chamber design showing the test seal in front of a dead-end section of tunnel excavation, the support steel surrounding the seal for simulating hitching, and the pressurization system using high-pressure water, compressed air, or some combination of the two. Sapko et al. [2005] describe the large and small hydrostatic test chambers in the LLEM in greater detail.

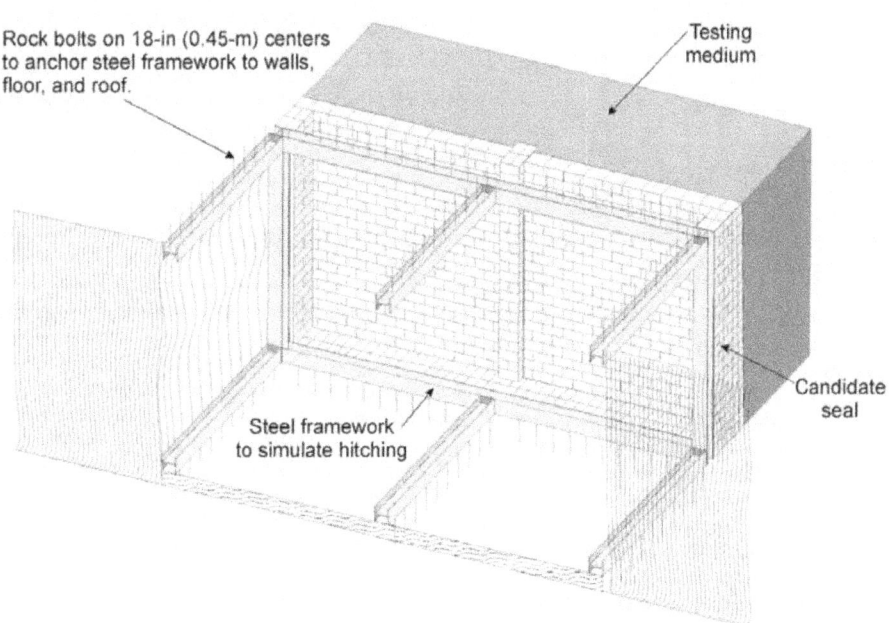

Figure 3.—Schematic of the hydrostatic chamber.

INSTRUMENTATION AND DATA COLLECTION

Pressure Waves From Test Explosions

Upon ignition with the electric matches, the methane-air mixture in the gas zone begins to burn and the flame front accelerates rapidly. In some tests, NIOSH researchers placed water-filled 55-gal barrels in the gas zone to create additional turbulence, which accelerates the flame front more rapidly. In all seal-related tests, the explosion is characterized as a deflagration as opposed to a detonation, since the maximum flame speed never exceeds about 1,100 ft/s, which is the local sound speed for the unreacted methane-air mixture.

The accelerating flame front produces pressure waves that travel at the local sound speed ahead of the flame front and are characterized by a static pressure component and a dynamic or velocity component. Both of these pressure components are time-dependent. In the free flow field, the sum of these two pressure components is the total pressure, which is also time-dependent.

Initially, the pressure waves emanating from a methane-air explosion rise from initial static pressure to peak static pressure slowly over a period of many tens of milliseconds. The time to go from initial to peak static pressure is termed the "rise time." The leading and lower pressure part of the blast wave travels at the local sound speed. However, at higher pressure, the local sound speed increases slightly. As the pressure wave propagates, the lagging higher pressure part of the wave, which is traveling slightly faster than the leading lower pressure part of the wave, will gain on the leading lower pressure edge of the wave. Via this mechanism, the rise time of the pressure wave will decrease and the blast wave may develop into a shock wave with instantaneous rise time.

In practice, the blast waves created during some seal-related explosion tests in the LLEM had rise times on the order of 10 ms. Theoretical relationships for shock waves with instantaneous rise times apply satisfactorily to blast waves with finite rise times of this magnitude.

The dynamic pressure p_V at the shock front is related to the static overpressure p_S by [Glasstone and Dolan 1977; Kinney 1962; Landau and Lifshitz 1987; Zucrow and Hoffman 1985]:

$$p_V = \left(\frac{5}{2}\right)\left(\frac{p_S^2}{7 p_o + p_S}\right) \tag{1}$$

where p_o = initial pressure.

For weak shock waves where p_S goes to zero, p_V also goes to zero; for strong shock waves where p_S becomes large, p_V also becomes large.

When a shock wave strikes a surface such as a seal head on, reflected overpressure p_R on the seal is given by [Glasstone and Dolan 1977; Kinney 1962; Landau and Lifshitz 1987; Zucrow and Hoffman 1985]:

$$p_R = 2 p_S \left(\frac{7 p_o + 4 p_S}{7 p_o + p_S}\right) \tag{2}$$

where p_S = static overpressure,
and p_o = initial pressure.

Equation 2 applies to a nonreactive shock or blast wave in which no chemical reactions are occurring when the reflecting surface is struck. For weak shock waves where p_S approaches zero, the reflected overpressure p_R is two times the incoming static overpressure p_S. For strong shock waves where p_S becomes large, the reflected overpressure p_R can be up to eight times the incoming static overpressure. For example, if the static overpressure is 117 psig, then the reflected overpressure is about 595 psig, or about five times the incoming static overpressure.

Pressure and Displacement Measurement Locations

NIOSH researchers recorded the P-t history applied to seals using different instrumentation arrangements that evolved during the course of the 20-psi seal testing program. Figure 4 shows the different pressure measurement locations with respect to the tested seals. In all tests, pressure transducers in the data-gathering panels located on the rib of each drift in the LLEM recorded the static pressure of the blast waves. In some tests, supplemental instrument stations located in the middle of the test drift contained pressure transducers to record the total pressure of the blast waves in the free field.

Additional instrument stations located in front of test seals in the crosscuts recorded static pressure of the passing blast waves (Figure 4). Test seals constructed in the crosscuts were recessed from the main drift anywhere from 0 to 8 ft. These supplemental instrument stations were placed anywhere from 1 to 4 ft in front of these crosscut seals. The number of instrument stations in front of each seal also varied from one to three, with instruments located either at the seal centerline, along the sides, or in combination. Again, the instrument stations in front of these test seals located in crosscuts generally recorded the static pressure of the passing blast wave. However, the measurements could also contain some of the dynamic component of the pressure wave as well as reflected pressure from wave impact on crosscut ribs facing the direction of wave propagation.

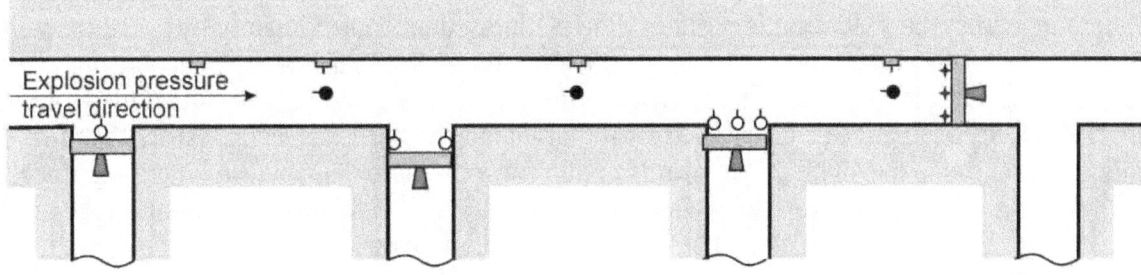

Figure 4.—Schematic of pressure and displacement measurement points for typical explosion tests in the LLEM.

Also shown in Figure 4 are instrument stations located directly in front of test seals in the main test drift. These instruments, located about 1 ft in front of the test seal, recorded the reflected wave overpressure on the test seal. The number of instrument stations varied, and the stations were located either at the seal centerline, along the sides, or in combination.

Figure 4 also shows the location of the LVDT to measure the displacement response of the structure. When used during a test, the LVDTs were located at midheight along the seal centerline.

Loading Conditions for Seal Tests

NIOSH researchers used four distinct test procedures for conducting tests on 20-psi seals. Each procedure subjected the test seal to different loading conditions as follows:

1. Explosion tests on seals in crosscuts loaded seals with the static blast wave overpressure that is nonuniform across the tested seal face.

2. Explosion tests on seals in C-drift loaded seals with the reflected blast wave overpressure that is assumed uniform across the seal.

3. Hydrostatic chamber tests using water pressure loaded seals with a static pressure that is nearly uniform across the seal except for the minor gravity effect.

4. Hydrostatic chamber tests using methane ignition pressure loaded seals with a static overpressure that is assumed uniform across the seal.

Most of the structural tests on seals described herein are of the first type—explosion tests on seals in crosscuts. The test explosion began at the closed end of C-drift, and the blast wave propagated down the drift and loaded seals located in crosscuts perpendicular to the direction of the main blast wave. That test procedure induced a nonuniform, static pressure that swept across the seal face as the pressure wave propagated down the entry. The rise times of the loading are longer than the natural period of the seals, so for structural purposes, the loading is considered static. The test seal could also experience some of the dynamic pressure component and the effects of turbulence depending on how far into the crosscut the test seal is recessed. At this time, these effects are unknown and are assumed negligible.

The blast wave propagates at the local sound speed of about 1,100 ft/s, and since the seal has a width of about 20 ft, the blast wave traverses past the seal in about 18 ms. The nonuniform, sweeping static pressure across the seal at some point in time is represented approximately by an 18-ms window from the P-t curve of the blast wave. The magnitude, duration, and shape of the blast wave varied considerably from test to test. Thus, the static pressure on the seal could vary significantly from the upstream to downstream edge of the seal. For structural analysis purposes, each test will require evaluation to determine the significance of this nonuniform pressure distribution on the seal face.

Several recent structural tests on seals were of the second type—explosion tests on seals in C-drift. The test explosion began at the closed end of C-drift, and the blast wave loaded seals located across C-drift. This test procedure induced a reflected blast wave overpressure on the seal face that is assumed uniform.

The hydrostatic chamber tests using either water pressure (type 3 test) or methane ignition pressure (type 4 test) applied a static pressure across the seal face. In both test procedures, the pressure on the seal face is assumed uniform. For the water pressure test (type 3), the gravity component of the water pressure behind the test seal is assumed negligible. In the case of the methane ignition tests, the elapsed time (rise time) to develop the pressure is very long with respect to the natural period of the tested seals. Therefore, the loading is considered static for structural analysis.

Boundary Conditions for Seal Tests

The roof, rib, and floor rocks in the LLEM are limestone with a compressive strength of about 24,200 psi and a modulus of elasticity of about 9,600,000 psi, based on laboratory tests conducted by Dolinar [2008]. The rock mass in the LLEM using the 1989 version of the Rock Mass Rating (RMR) system is a good-quality rock mass, with an RMR ranging from 77 to 79 according to Esterhuizen [2008]. Therefore, the foundation conditions for seal structures constructed and tested in the LLEM represent best-case circumstances and can be described as "rigid" or "unyielding." The foundation conditions for the seal tests in the LLEM as reported here do not represent typical conditions found in underground coal mines where the roof and floor rock and the coal ribs may have lower stiffness and strength.

Response Times, Time Constants, and Frequency Responses for Sensors Used in the LLEM

The response of a sensor to a step input depends on the range of the sensor and the time constant or frequency response of the sensor as given by the following relations:

$$\frac{P}{P_C} = 1 - \exp\left(\frac{-t}{T}\right) \qquad (3)$$

$$T = 1/F \qquad (4)$$

where P = magnitude of a step input,
P_C = full-scale range of sensor,
t = response time of sensor,
T = time constant of sensor,
and F = frequency response of sensor.

Figure 5 shows a plot of Equation 3 in dimensionless form. For an instantaneous unit step input ($P/P_C = 1$), an electronic sensor will require about three dimensionless time units to reach 95% of the step input.

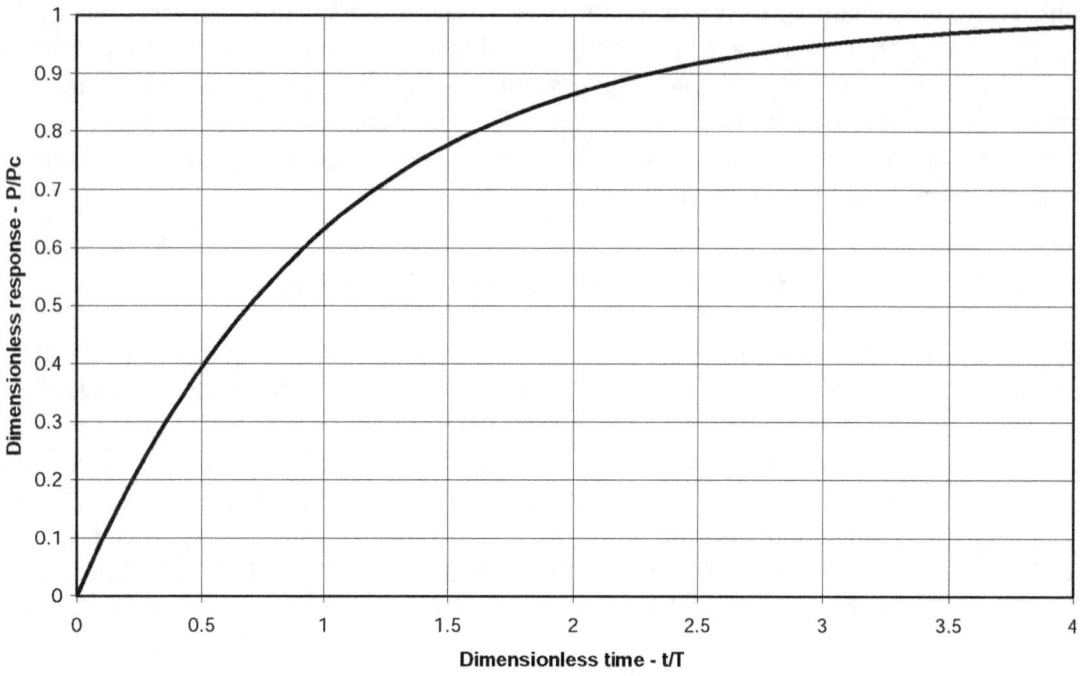

Figure 5.—Dimensionless response function for instantaneous unit step input.

Given the measured response time of a sensor to some fraction of the full-scale range of the sensor, the time constant and frequency response of the instrument are derived from Equations 3 and 4 as follows:

$$T = t/\ln\left(1 - \frac{P}{P_C}\right) \tag{5}$$

$$F = -\ln\left(1 - \frac{P}{P_C}\right)/t \tag{6}$$

NIOSH researchers used numerous brands of pressure transducers during the 20-psi seal testing program, including Patriot (AMETEK APT, Clawson, MI), Viatran (Viatran Corp., Grand Island, NY), and Transmetrics (Trans Metrics, Division of United Electric Controls, Watertown, MA). All transducers used a strain-gauge array that is bonded to a flat diaphragm to measure pressure-induced deflections of the diaphragm. The capacity of the pressure transducers ranged from 50 to 300 psi depending on requirements of a particular explosion test. Table 3 presents the response time (t) provided by Transmetrics, the manufacturer for the pressure transducers used in the most recent explosion tests, and the calculated time constant (T) and frequency response (F). In the calculations, it is assumed that the response time (t) is measured at 90% of the full-scale range of the sensor, i.e., $P/P_C = 0.9$.

Table 3.—Response time, time constant, and frequency response for various pressure transducers used to record P-t data during LLEM tests

Full-scale range of sensor (psi)	Measured response time t at 90% of full-scale (ms)	Calculated time constant T (ms)	Calculated frequency response F (Hz)
50	<1.6	0.70	1,439
100	<1.0	0.43	2,300
200	<0.7	0.30	3,290
300	<0.37	0.16	6,220

The data in Table 3 provide a means to judge whether pressure data measured with a particular instrument accurately record the actual phenomena. Figure 6 presents the calculated P-t response function for a 100-psi step input using a 100-psi transducer with a 1.0-ms response time. As indicated by Figure 6, the 100-psi transducer subject to a 100-psi step input will develop 95% of its response in about 1.4 ms. When subject to a smaller step input, the transducer will respond according to the time function shown in Figure 6. Using a 100-psi pressure transducer, which is typical for most of the data presented herein, the transducer will record the actual phenomena to within 10% of the actual pressure as long as the rise time is more than about 1.4 ms. Therefore, when a data point defined by the rise time of the recorded response and the magnitude of the peak pressure lies below the response function shown in Figure 6, the transducer will record the phenomena reliably.

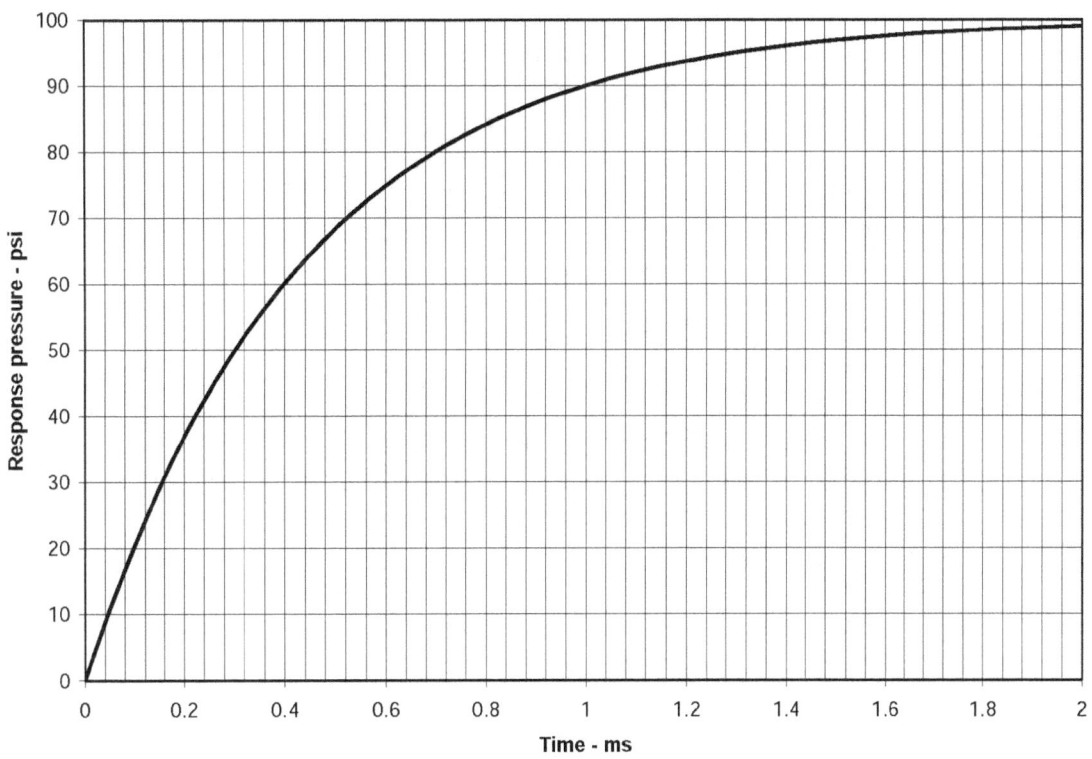

Figure 6.—Calculated pressure response for instantaneous 100-psi step input. Pressure transducer has 1.0-ms response time to 90% of full-scale and 2,300-Hz frequency response.

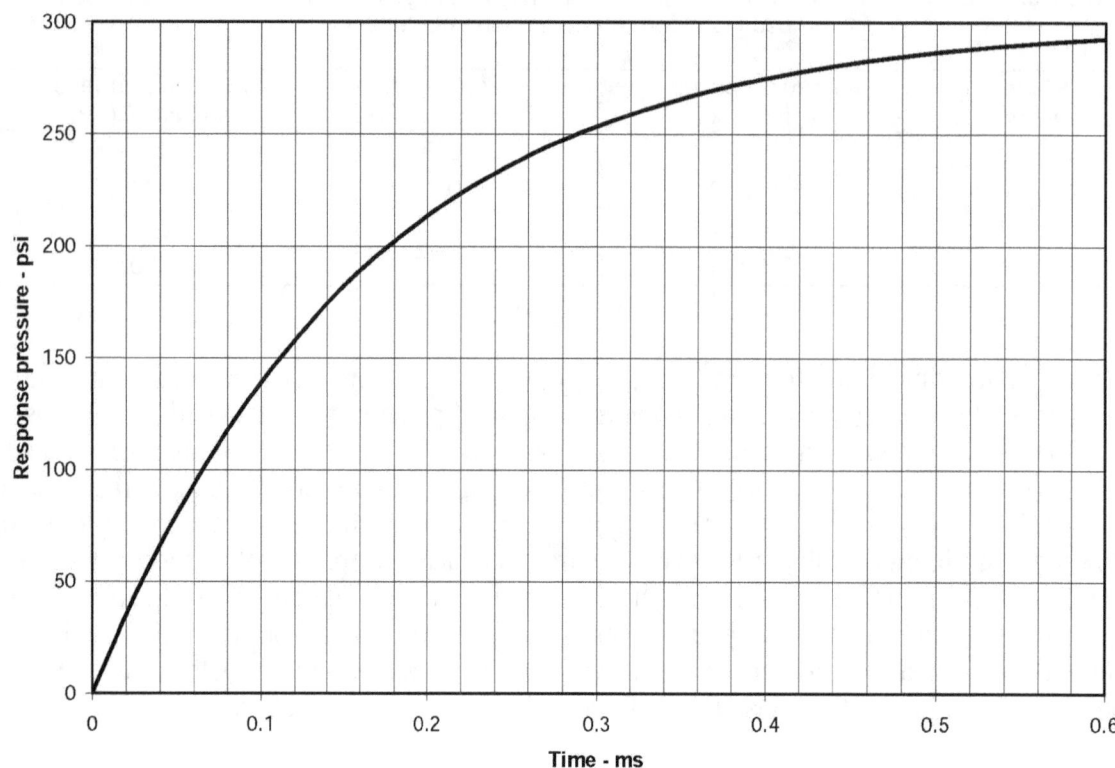

Figure 7.—Calculated pressure response for instantaneous 300-psi step input. Pressure transducer has 0.37-ms response time to 90% of full-scale and 6,220-Hz frequency response.

Figure 7 presents the calculated P-t response function for a 300-psi step input using a 300-psi transducer with a 0.37-ms response time. As indicated in Figure 7, a 300-psi transducer will develop 95% of its response in about 0.4 ms. With a 300-psi pressure transducer, which was used in some of the more recent, higher-pressure explosion tests in the LLEM, the transducer will record the actual phenomena to within 10% of the actual pressure as long as the rise time is more than about 0.4 ms. The recorded P-t curve is acceptable if a data point defined by the rise time of the recorded response and the magnitude of the peak pressure lies below the response function shown in Figure 7.

NIOSH researchers closely examined all of the P-t curves presented in the appendix to this report to confirm that the data meet the response time criteria presented by the information in Table 3 and the response time functions shown in Figures 6 and 7. In general, the initial rise times are much greater than 10 ms, which is well within the response capabilities of the pressure transducers used in these experiments. A point defined by the measured rise time and the magnitude of the peak pressure always lies below the calculated response function, indicating that the data quality is acceptable and that the measured pressure data accurately reflect the actual pressure developed during the test.

NIOSH researchers used LVDTs manufactured by Honeywell Sensotec (Columbus, OH) to measure displacement response of a seal at its centerline. Measurement range for these instruments is up to 6 in. The frequency response for these instruments as stated by the manufacturer is 300 Hz, which applies when the measurement rod of the displacement transducer is coupled directly to the structure. As will be discussed later, this method of coupling was generally not done in the D-t test data reported herein. Table 4 presents the frequency response provided by the manufacturer and the calculated time constant and response time assumed at 95% of the full-scale range of the sensor, i.e., $P/P_c = 0.95$. Figure 8 presents the calculated D-t response to a 6-in step input using a 6-in transducer with a 10-ms response time. The measured D-t data will reliably reflect the actual D-t phenomena when the rise time up to full-scale exceeds 10 ms. For the few tests presented here where the displacement transducer was coupled directly to the structure, the measured rise times exceed 10 ms and the recorded D-t curves are therefore considered reliable.

Table 4.—Response time, time constant, and frequency response for displacement transducers used to record D-t data during LLEM tests

Full-scale range of sensor (in)	Response time t at 95% of full-scale (ms)	Calculated time constant T (ms)	Frequency response F (Hz)
6 with direct coupling	<10 (calculated)	3.33	300 (from manufacturer)
6 with nylon filament coupling	<75 (measured in LLEM)	25.0	40 (calculated)

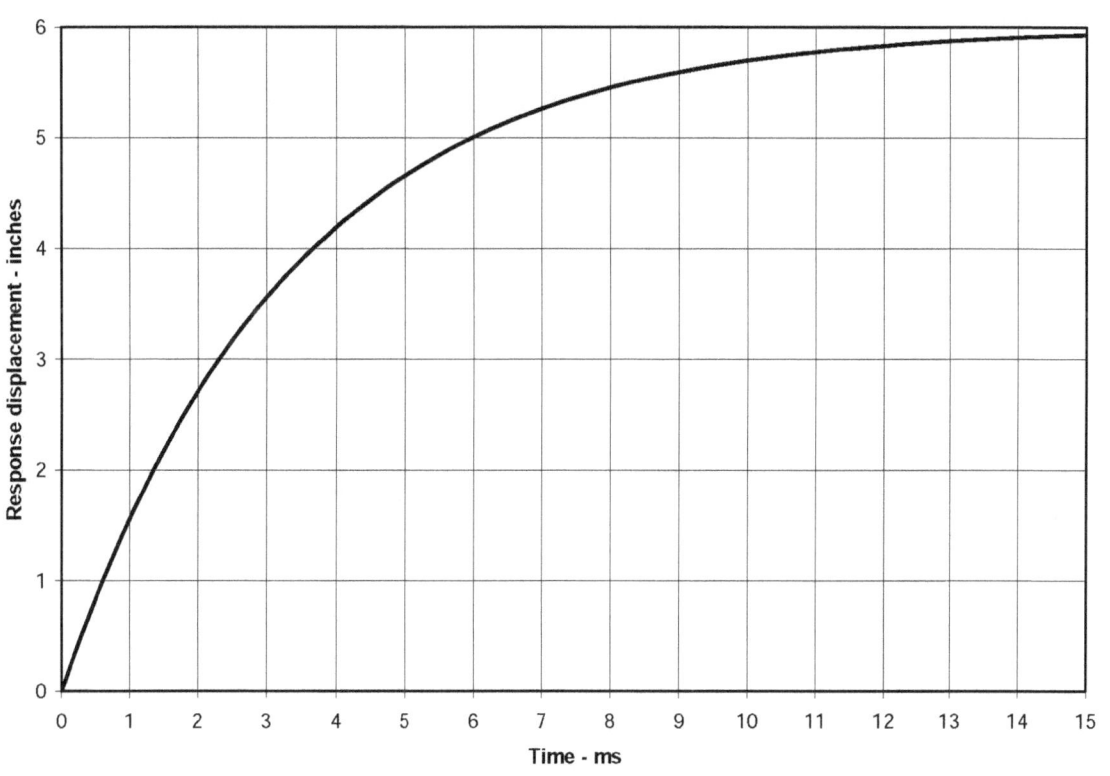

Figure 8.—Calculated displacement response for instantaneous 6-in step input. Displacement transducer is coupled directly to the structure and has 10-ms response time to 95% of full-scale and 300-Hz frequency response.

NIOSH researchers also employed another method to couple the displacement transducer to the structure using a spring-loaded nylon filament to protect the LVDT from damage during an explosion test. NIOSH researchers measured the full-scale response time of the LVDT using this experimental arrangement as about 75 ms. Based on this measured response time, the calculated time constant and frequency response for the LVDT are shown in Table 4. Figure 9 presents the calculated D-t response to a 6-in step input using a 6-in transducer with a 75-ms response time. The measured D-t data will reliably reflect the actual D-t phenomena when the rise time up to full-scale exceeds 75 ms. For all of the tests presented herein where the displacement transducer was coupled to the structure with a spring-loaded nylon filament, the measured rise times exceed 75 ms and the recorded D-t curves are therefore considered reliable.

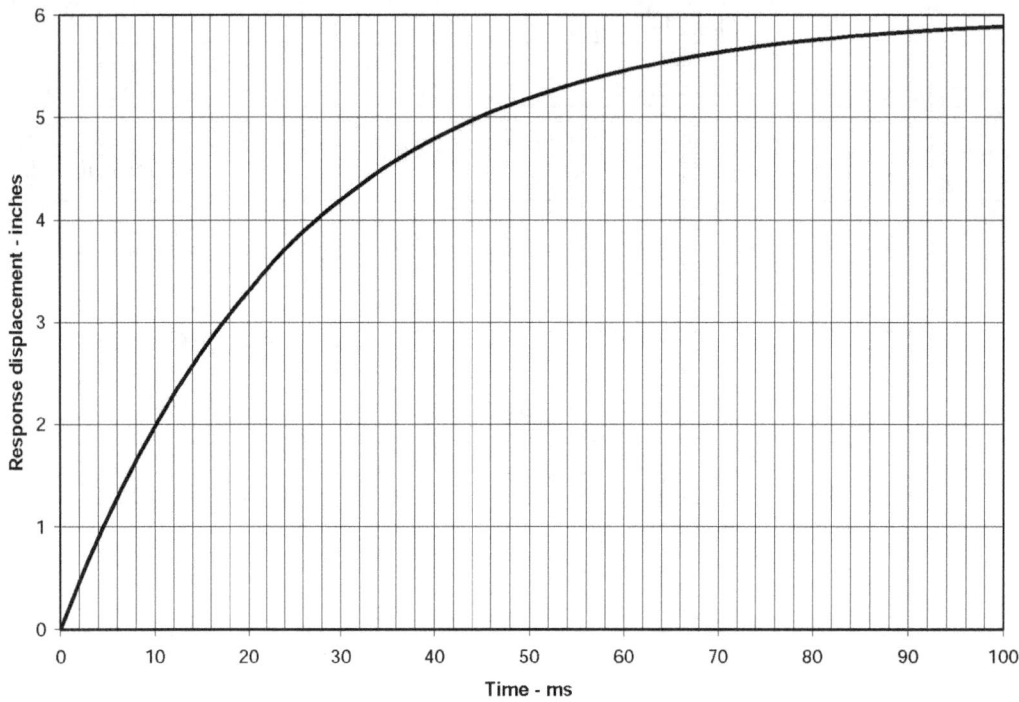

Figure 9.—Calculated displacement response for instantaneous 6-in step input. Displacement transducer is coupled via nylon filament to the structure and has 75-ms response time to 95% of full-scale and 40-Hz frequency response.

Data Acquisition System Characteristics

A PC-based data acquisition system by National Instruments Corp. (NI) (Austin, TX) recorded data from the various instruments at a sampling rate of 1,500 per second. In recent tests, a data acquisition system by Kinetic Systems (KS) (Boston, MA) also recorded data from the various instruments in parallel at a sampling rate of 5,000 per second. For consistency, all data reported here come from the NI system operating at 1,500 samples per second.

Quality of Pressure-Time and Displacement-Time Measurements

Figure 10 presents the P-t and D-t data for the entire duration of an explosion test in the LLEM. Beyond the initial arrival and decay of the first pressure wave, the data exhibit noise spikes and other extraneous features caused by a variety of experimental realities. Two questions arise with any experimental data:

1. Are the data real and representative of the physical phenomena under study?
2. Are the transducers and data acquisition system adequate to record the phenomena with sufficient detail for analysis?

To answer these questions, Figures 11–12 present a close examination of P-t data from a recent LLEM test that generated a reflected (head-on) explosion pressure of about 200 psi; Figure 13 presents the D-t data.

Figure 11 presents P-t curves from two separate pressure transducers located close together using an expanded time scale centered near the arrival of the first pressure wave. These transducers were located on the upstream side of a seal installed across the explosion entry. The green curve is an expanded view of the same P-t curve shown in Figure 10. The NI data acquisition system recorded the data at 1,500 samples per second. By inspection, both pressure transducers record similar signals with similar peak pressures of about 200 psi and similar rise times to peak pressure of about 3–4 ms. The observed pressure rise of 200 psi over a rise time of about 3–4 ms is within the capability of the instrument, as shown in Figure 6. NIOSH researchers conclude that the observed peak pressures and rise times represent the pressure waves under study.

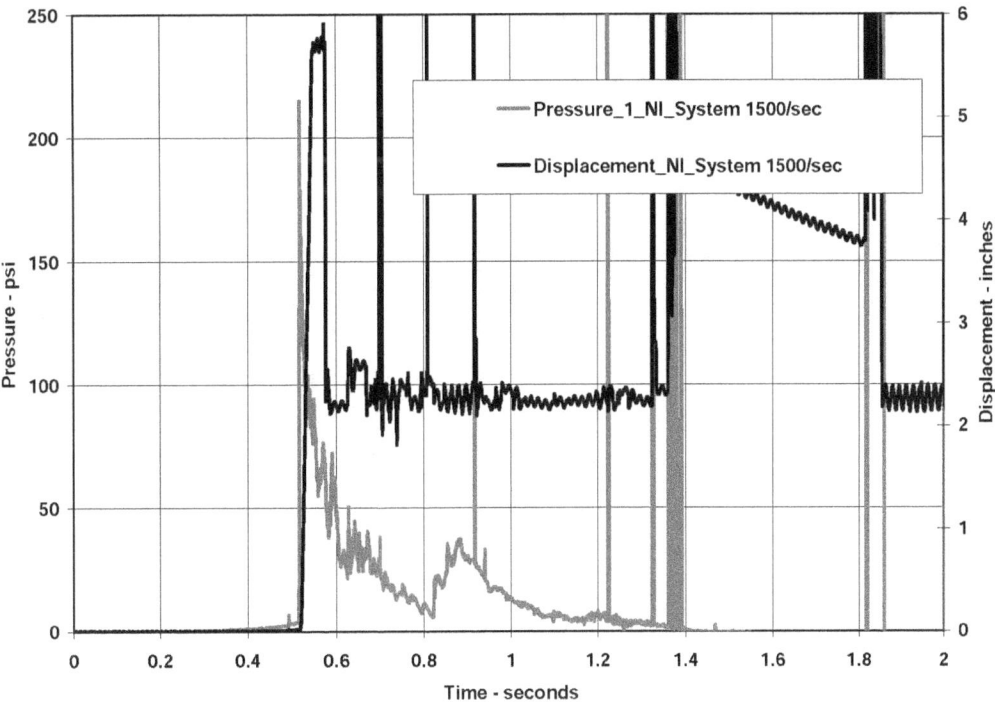

Figure 10.—Typical P-t and D-t data for the entire duration of an explosion test in the LLEM. Note noise spikes in data beyond 0.7 sec.

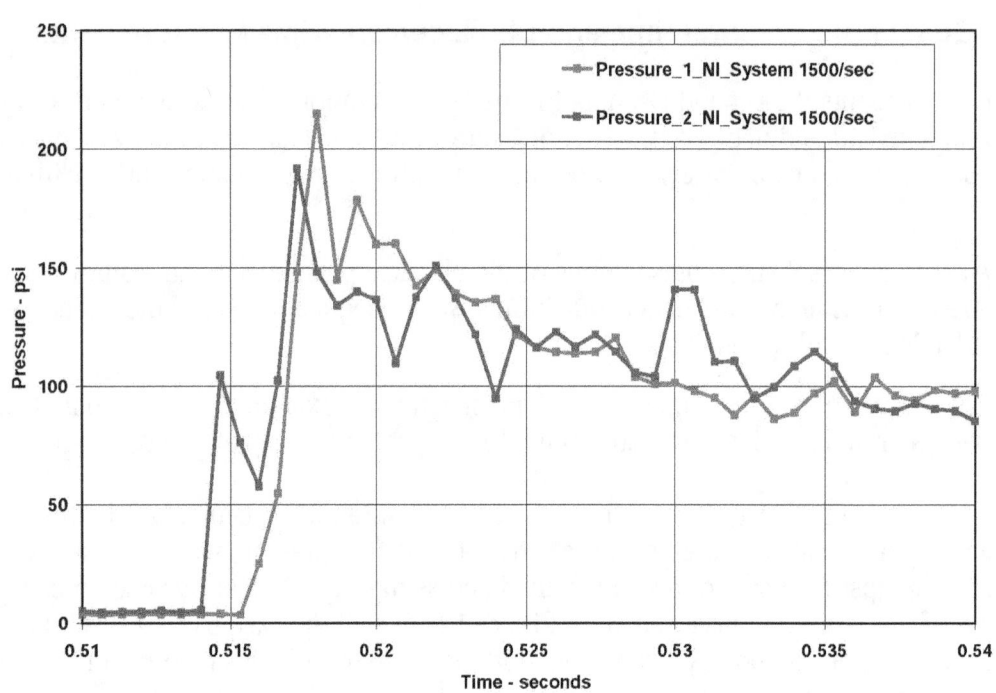

Figure 11.—Expanded time-scale view of P-t data from two separate pressure transducers. Data are recorded with the NI system at 1,500 samples per second. The green P-t curve is the same P-t curve shown in Figure 10.

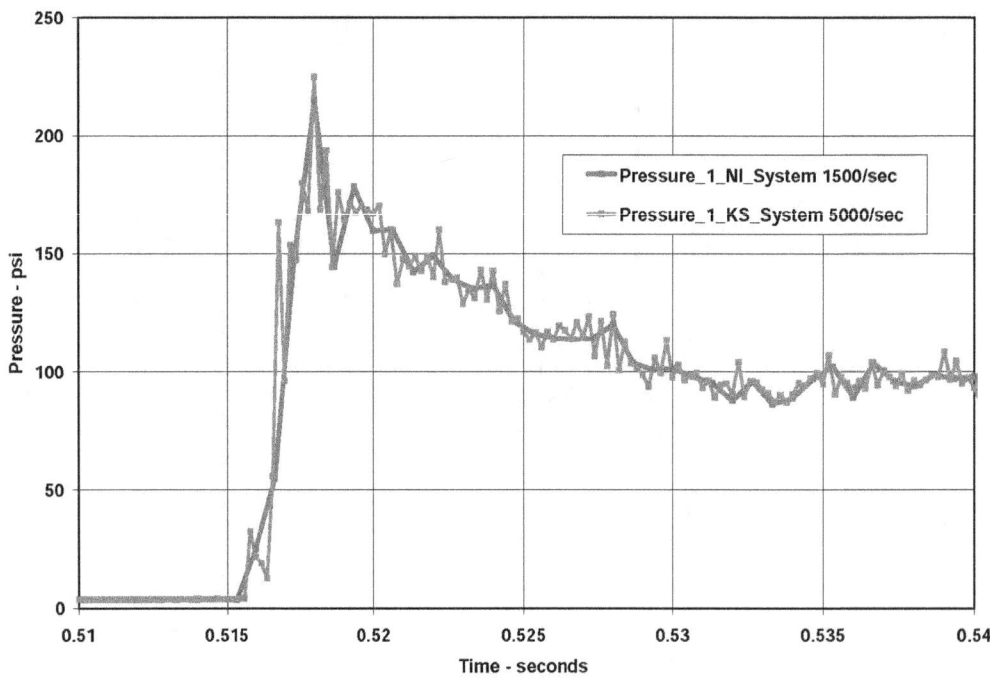

Figure 12.—Expanded time-scale view of P-t data from the same pressure transducer recorded with separate data acquisition systems operating at 1,500 and 5,000 samples per second.

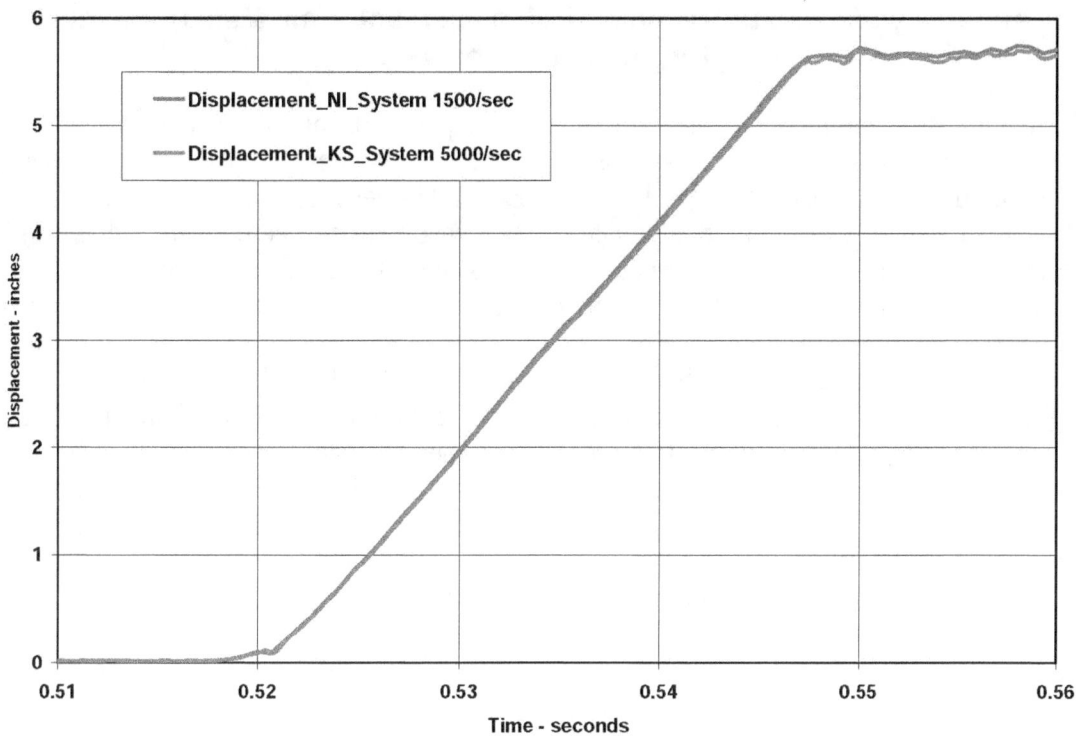

Figure 13.—Expanded time-scale view of D-t data from the same displacement transducer recorded with separate data acquisition systems operating at 1,500 and 5,000 samples per second.

Figure 12 presents an expanded view of the same P-t curve shown in Figure 10, but as recorded with two separate data acquisition systems operating at 1,500 samples per second (the NI system) and 5,000 samples per second (the KS system). By inspection, both recording systems captured similar signals, although the faster KS system provides more detail. NIOSH researchers conclude that the pressure transducers and data acquisition system are adequate to record the P-t curves from these explosion tests.

Figure 13 presents an expanded view of the D-t curve shown in Figure 10 as recorded with two separate data acquisition systems operating at 1,500 and 5,000 samples per second. For a displacement transducer coupled directly to the structure, the observed displacement of about 5.7 in over a time of 25 ms is within the capability of the instrument, as shown in Figure 8. NIOSH researchers conclude that the displacement transducer and data acquisition system are well matched to record the D-t curves from these explosion tests. However, NIOSH researchers note that while the system can measure the initial displacement response of the seal structures, it may not measure the later vibratory response of the structures because of the technique used to couple the LVDT to the structure. Thus, the displacement data are only reliable up to the initial peak displacement and not beyond that point in time.

Adequacy of Pressure-Time and Displacement-Time Measurements for Structural Analysis

A final question about the P-t data concerns its adequacy for reliable structural analysis. As demonstrated in the previous sections, the pressure transducers and data acquisition system adequately capture pressure data with a rise time of at least 100 psi/ms. Higher-frequency components may exist in the P-t data that may not be captured adequately. The question arises as to whether these higher-frequency components are important in structural analysis of seals. NIOSH researchers performed a simple structural analysis aimed at discerning the possible importance of these higher-frequency P-t components.

For this structural analysis, NIOSH researchers used the Wall Analysis Code (WAC) from the U.S. Army Corps of Engineers [Slawson 1995]. WAC is a single-degree-of-freedom structural dynamics model that solves the equation of motion to determine the D-t history at midheight of a wall given some P-t curve applied to the wall. The analysis considered a simply supported, 2-ft-thick concrete wall. Figure 12 shows the P-t curves considered in the analysis, which were measured with the NI and KS data acquisition systems at sampling rates of 1,500 and 5,000 samples per second, respectively. Presumably, the data set collected with the KS system contains more of the higher-frequency components of the P-t curve.

Figure 14 shows the computed D-t responses from these two different P-t curves. The different responses are purposely offset in time for clarity. As seen by inspection, no significant difference exists between the two computed D-t responses. NIOSH researchers conclude that the P-t data collected by the NI system at 1,500 samples per second are adequate for structural analysis of seals under the conditions of the explosion tests conducted in the LLEM for evaluating 20-psi seals.

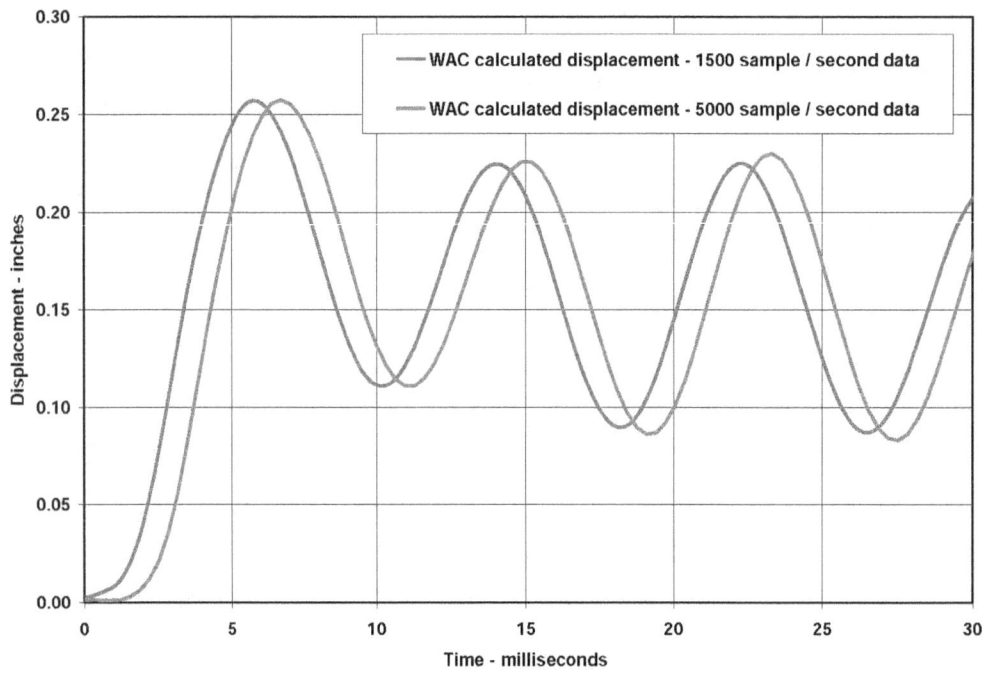

Figure 14.—Computed D-t responses for hypothetical structure using P-t curves in Figure 12 as input for structural analysis. Calculated displacements are offset in time for clarity.

Comments on Smoothing

The P-t data reported here are *raw data*. All data reported came from the NI system in the LLEM, which records data at 1,500 samples per second for 5 sec. In prior NIOSH publications, the P-t data reported were *smoothed* data derived from the raw data using a 15-point moving-average algorithm. In this simple algorithm, the midpoint (point 8) in a 15-point data set is replaced with the average value of those 15 data points. Each data point in the moving 15-point set is given equal weight in the averaging. This smoothing algorithm tends to remove the highest peaks in the pressure data. It will also slow the rate of change in the pressure data, i.e., the rise time for fast pressure changes due to blast waves and other pressure transients will decrease. Although this smoothing algorithm can remove spurious data points due to noise or other transients in the data acquisition system, structural analysis requires the use of the raw data. The actual peak pressures and rise times measured are within the capabilities of the instruments and data acquisition system. True peak pressures and pressure-rise times are important when computing the response of a structure to some applied load. Because the data reported here are raw, there will be discrepancies with the smoothed (15 points, or over a 10-ms-wide moving window) peak pressures reported in previous NIOSH reports. The pressure data herein are always higher than data reported previously.

GENERAL CONSTRUCTION DETAILS FOR CATEGORY 1 THROUGH 6 SEALS

Category 1 Seals: Concrete or Concretelike Materials With Internal Steel Reinforcement and Anchorage to Rock

Category 1 seals are made of concrete or concretelike materials such as shotcrete or gunite. The seal structure contains internal steel reinforcement and is anchored to the surrounding rock with steel reinforcement bars. Two varieties of 20-psi seals are reported here—the Insteel 3-D seal from Precision Mine Repair, Inc., made with shotcrete and steel reinforcement bar, and the Meshblock seal from R. G. Johnson Co., Inc., made with pumped cement or shotcrete and steel reinforcement bar.

Category 1A seal: Insteel 3-D seal – shotcrete with reinforcement

The Insteel 3-D seal made by Precision Mine Repair, Inc., is constructed with shotcrete, steel reinforcement bar, and reinforcement wire. Steel reinforcement bars also anchor the seal to the surrounding rock. NIOSH researchers tested seven different structures using these concepts and materials during development of a seal that met the previous 20-psi seal design standard (Table 1). Test results from one of these structures were reported by Sapko et al. [2005].

Figures 15 and 16 show front-, plan-, and side-view drawings of the Insteel 3-D seal for the former 20-psi seal design standard. The seal consists of a three-dimensional welded wire space frame called an Insteel 3-D panel that is encased in concrete. The structural components of the seal are the rear Insteel 3-D panel with Stayform backing, the front Insteel 3-D panel without Stayform backing, #3 steel reinforcement bars laid horizontally from rib to rib within the Insteel 3-D panels, a plane of vertical #8 reinforcement bars and #8 reinforcement bar anchors into the roof and floor in front of each Insteel 3-D panel, and horizontal #8 reinforcement bar anchors into each rib. The panels are filled completely with fast-setting concrete mix applied from the front side of the seal using a shotcrete machine at a pressure of 100 psi. This mix hardens within 15 min and is nearly fully cured to design strength within 24 hr. The minimum design uniaxial compressive strength requirement of this shotcrete is 2,500 psi. The finished seal is at least 11.5 in thick.

To construct the seal, the rear Insteel 3-D panels are trimmed to fit opening dimensions at the installation site and laid horizontally across the opening. Each horizontal panel contains a #3 steel reinforcement bar laid horizontally from rib to rib and spaced less than 16 in apart. The rear panel contains Stayform backing on the inby side, which serves as a backing for spraying the shotcrete from the outby side. Figure 17 shows a closeup view of a rear Insteel 3-D panel with the Stayform backing.

Vertical holes are drilled into the roof and floor for two rows of 36-in-long, #8 steel reinforcement bar anchors. The holes are at least 12 in deep and evenly spaced across the entry on less than 24-in centers. The front and rear rows of holes are offset laterally from each other (Figure 15). The #8 steel reinforcement bar anchors in the rear row are epoxy-grouted in place, and vertical #8 reinforcement bars are tied in place using preformed clamps between corresponding roof and floor anchors. The rear Insteel 3-D panel with Stayform backing is then tied with wire to the rear row of vertical #8 steel reinforcement bar. Three holes are also drilled on 24-in centers into each rib at least 12 in deep for additional anchorage, and #8 steel reinforcement bar anchors are epoxy-grouted into place and tied to the #3 steel reinforcement bars within the Insteel 3-D panels. Figure 18 shows the reinforcement construction at this stage.

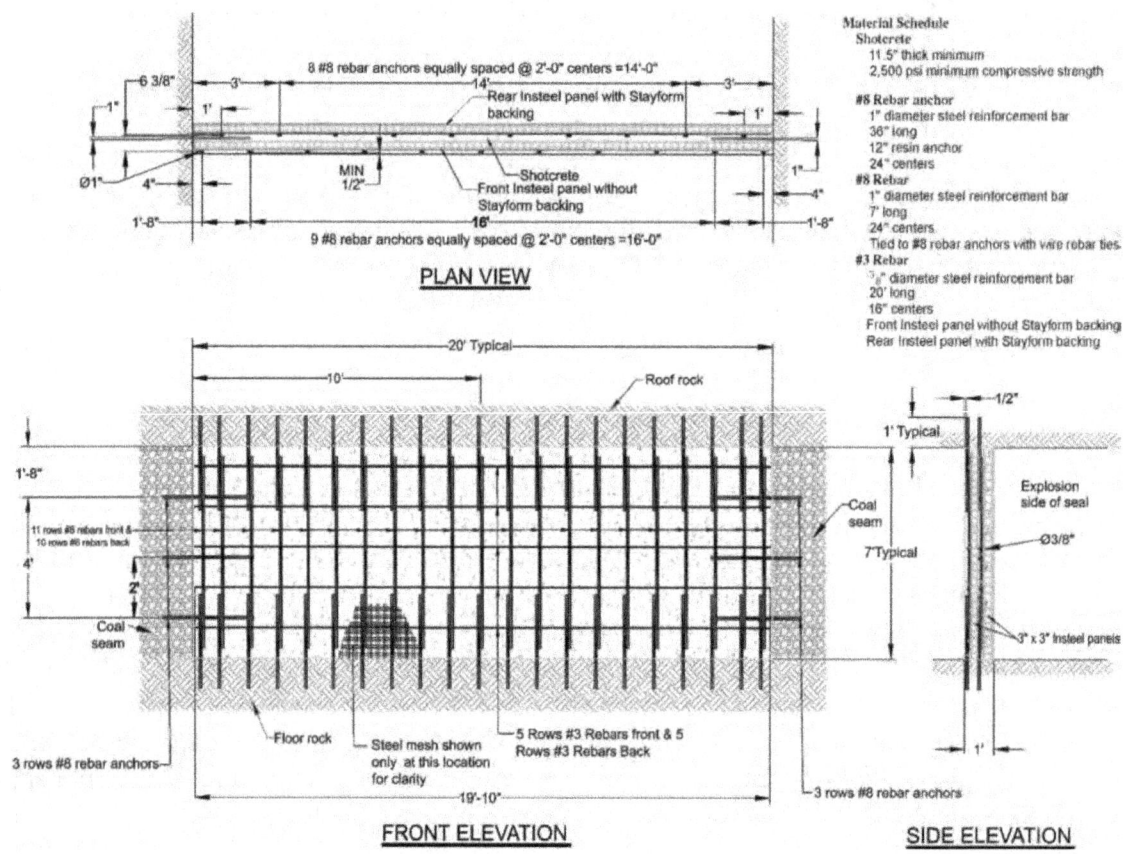

Figure 15.—Front-, plan-, and side-view drawings of Category 1A structure: Insteel 3-D seal from Precision Mine Repair, Inc.

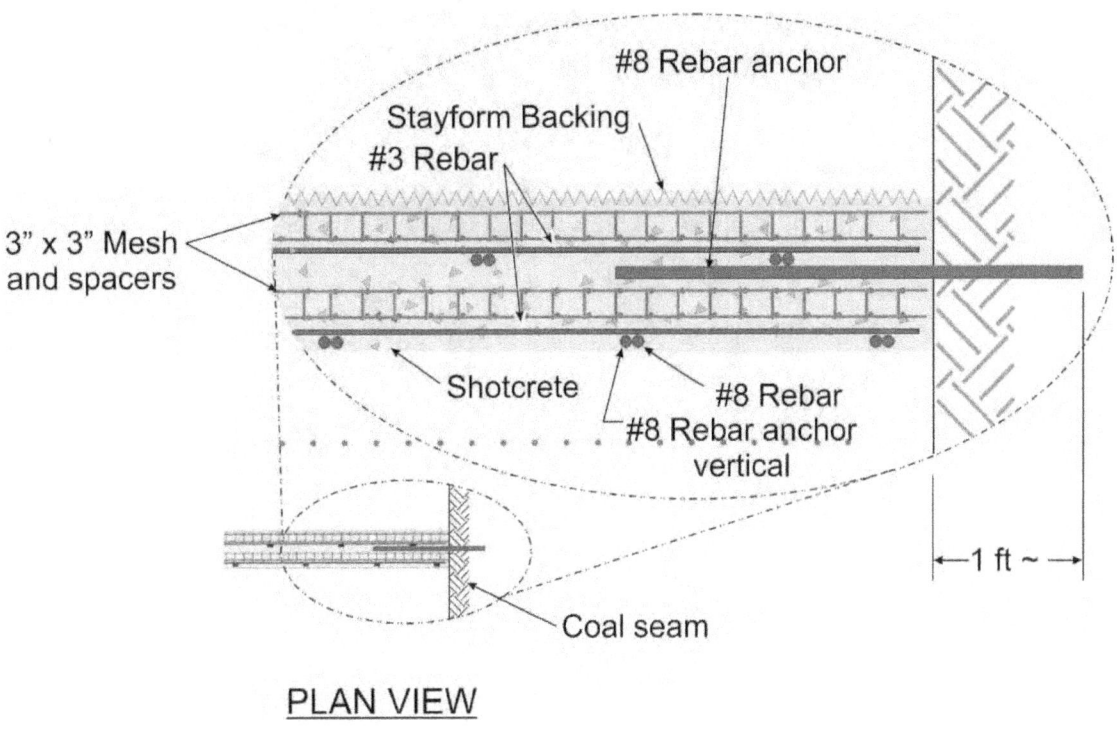

Figure 16.—Detailed plan-view drawing of Category 1A structure: Insteel 3-D seal from Precision Mine Repair, Inc.

Figure 17.—Closeup of a rear Insteel panel with Stayform backing and the horizontal #3 steel reinforcement bars.

Figure 18.—Insteel 3-D seal under construction showing the rear Insteel panel with Stayform backing, one plane of vertical #8 steel reinforcement bars and anchors, the horizontal #8 steel reinforcement bar anchors, and the horizontal #3 steel reinforcement bars.

The front Insteel 3-D panels without Stayform backing are then trimmed to fit the opening dimensions and laid horizontally in front of the rear plane of vertical #8 steel reinforcement bar. The front row of #8 steel reinforcement bar anchors is epoxy-grouted into place, and vertical #8 steel reinforcement bars are tied in place using preformed clamps between corresponding roof and floor anchors. The front Insteel 3-D panel is tied with wire to the front row of vertical #8 steel reinforcement bars. Figure 19 shows the reinforcement construction at this stage. The wire mesh and steel reinforcement bar network is now ready for spraying with shotcrete with a minimum design uniaxial compressive strength of 2,500 psi to a depth of at least 11.5 in.

Figure 19.— Insteel 3-D seal under construction with the addition of the front Insteel panel and the front plane of vertical #8 steel reinforcement bars.

Category 1B seal: Meshblock seal – shotcrete with reinforcement

The Meshblock seal made by R. G. Johnson Co., Inc., is constructed with shotcrete, steel reinforcement bar, and reinforcement wire. Steel reinforcement bars also anchor the seal to the surrounding rock. NIOSH researchers tested six different structures using these concepts and materials during development of a seal that met the previous 20-psi seal design standard (Table 1). Weiss et al. [1999] and Sapko et al. [2005] provide additional construction details and test results for seals in this category, along with some test results.

Figure 20 shows front-, plan-, and side-view drawings of the Meshblock for the previous 20-psi seal design standard. The seal consists of three-dimensional welded-wire Meshblocks that are pumped full with concrete. The major structural components of the seal are a plane of vertical #8 steel reinforcement bars and #8 steel reinforcement bar anchors that attach the seal to the roof, floor, and both ribs. Figure 21 shows part of a typical Meshblock, which is made from 6-in by 6-in welded-wire mesh overlain by metal hardware cloth. The diameter of the 6-in by 6-in welded wires is about 0.16 in, and the opening size in the hardware cloth is about 0.12 in. Overall dimensions for a Meshblock are 42 in wide by 18.25 in high by 12 in deep. The space between the Meshblocks is filled with shotcrete (Figure 22). The shotcrete brand used is Quikrete MB-500, which is a mixture of cement and minus 0.20-in aggregate and has a minimum design uniaxial compressive strength of 5,800 psi.

To construct the seal, vertical holes are drilled 24 in deep into the roof and floor on 24-in centers. Additional horizontal holes are drilled 24 in deep into ribs on 40-in centers. Steel reinforcement bars, 48 in long and 1 inch in diameter, are resin-anchored into each hole (Figure 20). Vertical #8 steel reinforcement bars that overlap the anchor bars by 24 in are tied using preformed clamps to the corresponding roof and floor anchors. This reinforcement plane is centered in the Meshblock formwork.

The Meshblock formwork is then pumped full with shotcrete. The metal hardware cloth facing on the Meshblocks allows the nozzleman to examine the shotcrete flowing into the formwork (Figure 22). Normally, two layers of Meshblock units are filled at a time. Care is required to limit pouring delays to less than 30 min to prevent cold joints in the finished structure. The top 6–12 in of the seal requires cutting and fitting the Meshblock units to create a back wall, which is filled in with the shotcrete using the spray nozzle. Finally, the perimeter is sprayed to stabilize and seal the surrounding strata.

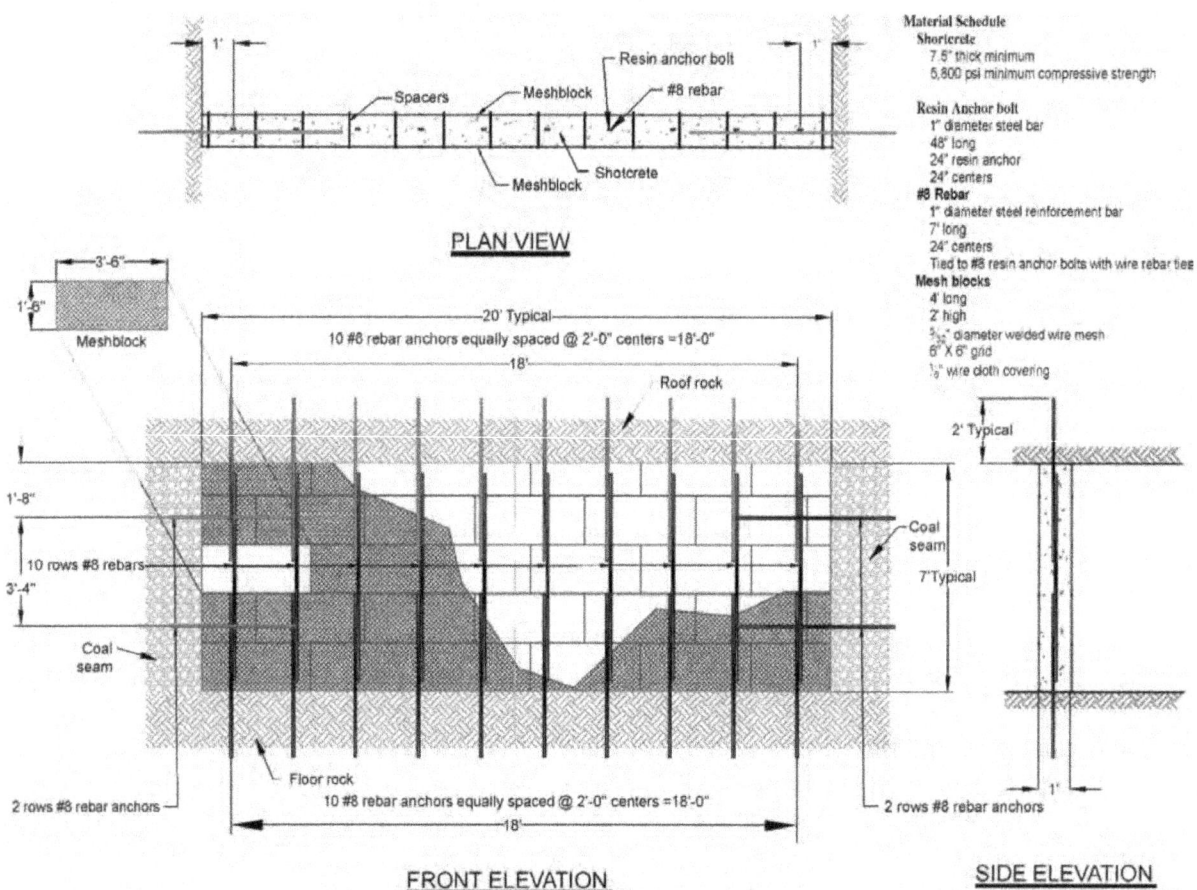

Figure 20.—Front-, plan-, and side-view drawings of Category 1B structure: Meshblock seal from R. G. Johnson Co., Inc.

Figure 21.—Closeup of Meshblock seal under construction showing the Meshblocks and the vertical #8 steel reinforcement bars and anchors.

Figure 22.—Meshblock seal under construction showing the lower Meshblocks, the vertical #8 steel reinforcement bars and anchors, and placement of shotcrete.

Category 2 Seals: Pumpable Cementitious Materials With No Steel Reinforcement and No Hitching

Category 2 seals are also called pumpable seals and are made from cementitious materials of various compositions. Different formulations will produce different compressive strengths for the pumpable cementitious material. Lower-strength materials will require a thicker seal. These seals do not contain any internal steel reinforcement and are not hitched to the surrounding rock. Friction between the seal material and the surrounding rock holds the seal in place. Three varieties of 20-psi seals are reported here with different compressive strength materials and different seal thicknesses.

Category 2A seal: Pumpable cementitious material with uniaxial compressive strength of 200 psi constructed 48 in thick

Category 2A seals include Tekseal manufactured by Minova, Inc., and Celuseal manufactured by R. G. Johnson Co., Inc. These seals are made from a pumpable cementitious material with a minimum design uniaxial compressive strength of at least 200 psi and a thickness of at least 48 in. NIOSH researchers tested four different structures in this category during development of a seal that met the previous 20-psi seal design standard (Table 1). Greninger et al. [1991], Weiss et al. [1993, 1996], Weiss and Harteis [2008], and Sapko et al. [2005] provide additional details on constructing the pumpable, cementitious material seals in this category along with some test results.

Figure 23 shows front-, plan-, and side-view drawings of the typical formwork for a pumpable cementitious material seal. The two form walls must have sufficient strength to resist the pressure from the uncured cementitious seal material. The formwork shown in Figures 23 and 24 consists of 4-in by 4-in vertical wood posts installed on 3-ft centers across the entry, 1-in by 6-in wood boards nailed horizontally across the timbers on less than 2-ft centers, and MSHA-approved brattice material nailed to the inside of the frame. The brattice material must overlap approximately 4 in and is secured onto each rib, the roof, and the floor. The forms do not add any significant structural strength to the seal. Finally, the volume between the two forms is pumped full of cementitious material (Figure 25).

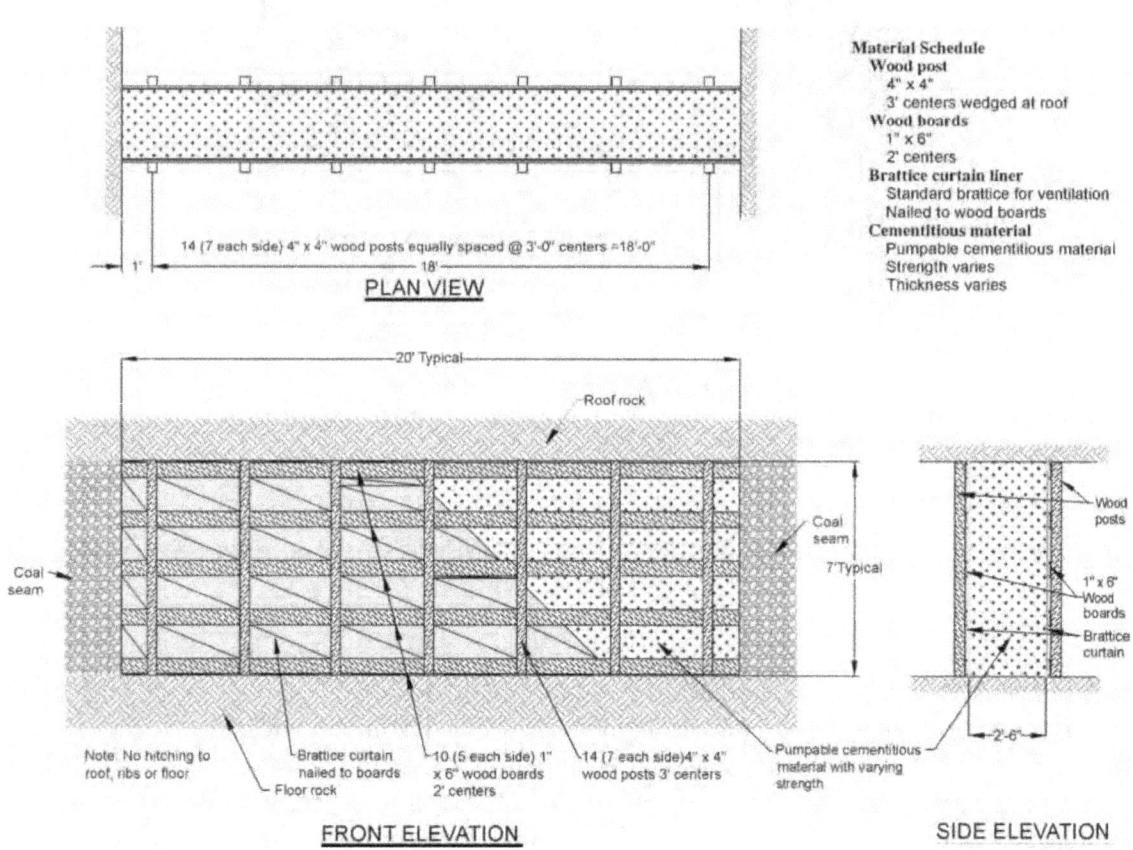

Figure 23.—Front-, plan-, and side-view drawings of Category 2 structure made from pumpable cementitious materials with no steel reinforcement and no hitching.

Figure 24.—Formwork for a typical pumpable seal showing the vertical posts, the horizontal boards, and the brattice liner (partially removed).

Figure 25.—Inside the formwork of a typical pumpable seal showing the brattice liner and the cementitious filling material.

Tekseal and Celuseal are both lightweight, noncombustible cement-based products. Cementitious powder, water, and air are metered into a continuous mixer and then pumped between the forms. The amount of cementitious material used per cubic yard of seal determines the density and strength of the seal material. These seals should normally be installed using a continuous pour during construction so that a solid plug with no cold joints is obtained.

Category 2B seal: Pumpable cementitious material with uniaxial compressive strength of 433 psi constructed 36 in thick

Category 2B seals include the Ribfill seal made by HeiTech Corp. Again, these seals are made from a pumpable cementitious material with a minimum design uniaxial compressive strength of at least 433 psi and a thickness of at least 36 in. NIOSH researchers tested one structure in this category during development of a seal that met the previous 20-psi seal design standard (Table 1). Weiss et al. [2002] provide additional details on constructing the pumpable, cementitious material seals in this category along with test results.

Construction of the seal uses a formwork similar to that shown in Figures 23 and 24, except that the distance between form walls is at least 36 in and the minimum design uniaxial compressive strength of the cementitious material must exceed 433 psi.

Category 2C seal: Pumpable cementitious material with uniaxial compressive strength of 480–600 psi constructed 24–30 in thick

Category 2C seals include the Rockfast seals also made by HeiTech Corp. NIOSH researchers tested four different structures in this category during development of a seal that met the previous 20-psi seal design standard (Table 1). Weiss et al. [2002] provide additional details on constructing the seals in this category along with test results.

The Rockfast M-FGL material used for these seals is a modified, portland cement-based, fiber-reinforced, pumpable grout that exhibits rapid gelation and high early strength. The 24-in-thick version of this seal requires material with a minimum design uniaxial compressive strength of 677 psi, while the 30-in-thick version requires material with a minimum design uniaxial compressive strength of 480 psi. Construction of the seal uses a formwork similar to that shown in Figures 23 and 24, except that the distance between form walls is 24–30 in.

Category 3 Seals: Articulated Structures – Solid and Hollow-Core Concrete Blocks With or Without Hitching

Category 3 seals are articulated structures made from discrete concrete blocks bonded together with a mortar. These seals may or may not be hitched into the surrounding rock, and they do not contain any internal steel reinforcement. Two varieties of seal structures are reported here: the standard solid-concrete-block seal, which is hitched into the surrounding rock, and a variant of the standard seal that uses pressurized grout bags in lieu of hitching to attach the seal structure to the surrounding rock through friction. Also included in this category are ventilation stoppings made from solid and hollow-core concrete blocks. These stoppings are constructed by dry stacking the blocks without mortar and without hitching and only wedging along the ribs and roof. Test results from these structures are included here since these structures can form an integral part of other seal structures in other seal categories.

Category 3A seal: Standard solid-concrete-block seal with hitching

The seal design in Category 3A is the original 20-psi standard seal, also known as the Mitchell-Barrett mine seal. The seal is constructed with wet-laid, solid concrete blocks, a central pilaster for added strength, and hitching into the ribs and floor. NIOSH researchers tested seven different structures in this category (Table 1). Greninger et al. [1989, 1991], Weiss et al. [1993, 1996] and Sapko et al. [2005] provide additional details on constructing the standard solid-concrete-block seals in this category along with test results.

Figure 26 shows front-, plan- and side-view drawings of the typical standard solid-concrete-block seal as constructed in the LLEM. The seal is constructed with solid concrete blocks measuring either 8-in by 8-in by 16-in or 6-in by 8-in by 16-in (laid flat) and placed with ASTM Type S mortar. The mortar is applied on all vertical and horizontal block joints to a nominal thickness of 3/8 in. The blocks in the first course are placed with their long axis parallel to the rib. Blocks in the second course are oriented perpendicular to the first course, and the front and back rows in this course may or may not be staggered.

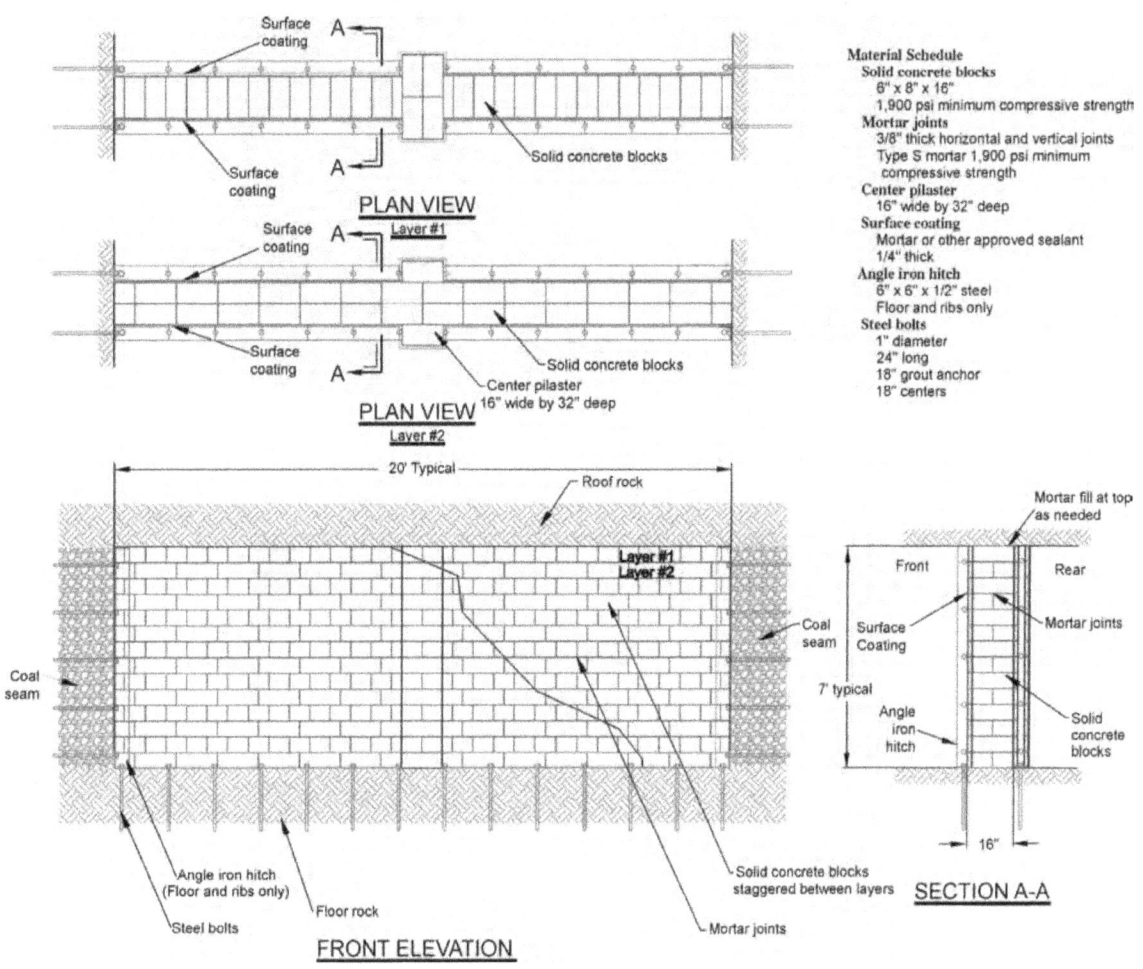

Figure 26.—Front-, plan-, and side-view drawings of Category 3A structure: standard solid-concrete-block seal with simulated hitching.

The seal is constructed 16 in thick with a center pilaster measuring 16 in wide by 32 in thick for added strength (Figures 26–27). Blocks are cut and laid on the last course so that a gap approximately 1–2 in high exists between the top course of block and the mine roof (Figure 28). This gap is filled completely with mortar across the width and depth of the seal. No wedges are used in the seal construction. Finally, the faces on both sides of the seal are coated with mortar or other surface sealant approved by MSHA.

In practice, the standard solid-concrete-block seal is hitched into the floor and rib by excavating at least 6 in into solid competent material. To simulate hitching during tests in the LLEM, NIOSH researchers use 6-in by 6-in by 0.5-in-thick steel angle (Figure 29), anchored on 18-in centers using 1-in-diam by 9-in-long anchor bolts. The nominal uniaxial compressive strength of the block material is 1,900 psi, although recent full-scale block testing by Barczak and Batchler [2008] indicates a solid-concrete-block uniaxial compressive strength of 1,330–1,780 psi. Design uniaxial compressive strength of the mortar exceeds 1,900 psi.

Figure 27.—Solid-concrete-block seal under construction showing the center pilaster and the fully mortared joints on all sides.

Figure 28.—In background, top of a solid-concrete-block seal showing small cut blocks and mortar filling at the seal top.

Figure 29.—Completed solid-concrete-block seal without a center pilaster showing the angle iron hitch on the ribs and floor only.

Category 3B seal: Solid-concrete-block seal with Packsetter Bags and without hitching

The 20-psi seal design in Category 3B is similar to the standard solid-concrete-block design described above, except that Packsetter Bags distributed by Strata Mine Services are used in place of conventional hitching. NIOSH researchers tested three different structures in this category (Table 1). Weiss et al. [2002] provide additional details on constructing the concrete block with grout bag seals in this category along with test results.

Figure 30 shows front-, plan-, and side-view drawings of the typical standard solid-concrete-block seal with Packsetter Bags as constructed in the LLEM. The seal is constructed with tongue-and-groove, solid concrete blocks measuring either 8 in by 8 in by 16 in or 6 in by 8 in by 16 in (laid flat) and placed with mortar on all surfaces to a nominal thickness of 3/8 in. The mortar used must meet ASTM C270-91a as Type N, S, or M mortar. BlocBond from Quikrete is also considered an acceptable mortar. The seal has a central pilaster measuring 16 in wide by 32 in deep.

Construction of this seal begins with placement of the first course of tongue-and-groove block laid parallel to the rib in wet mortar to within 2 in of each rib. A Packsetter Bag is laid flat and positioned at least 6 in under the outside corners of the first course of blocks. Placement of wet-laid concrete blocks continues until the blocks come within 2–5 in of the roof and ribs. Packsetter Bags placed along the ribs and across the roof overlap adjacent bags by a minimum of 6 in and overlap the front and back face of the seal by at least 3 in. The top-right and top-left corner Packsetter Bags are placed with half the bag down the rib side and half the bag across the roof side of the seal. Because of possible rib and roof undulations, if the distance between seal and strata is up to 8 in due to a localized cavity, a spacer Packsetter Bag can be used.

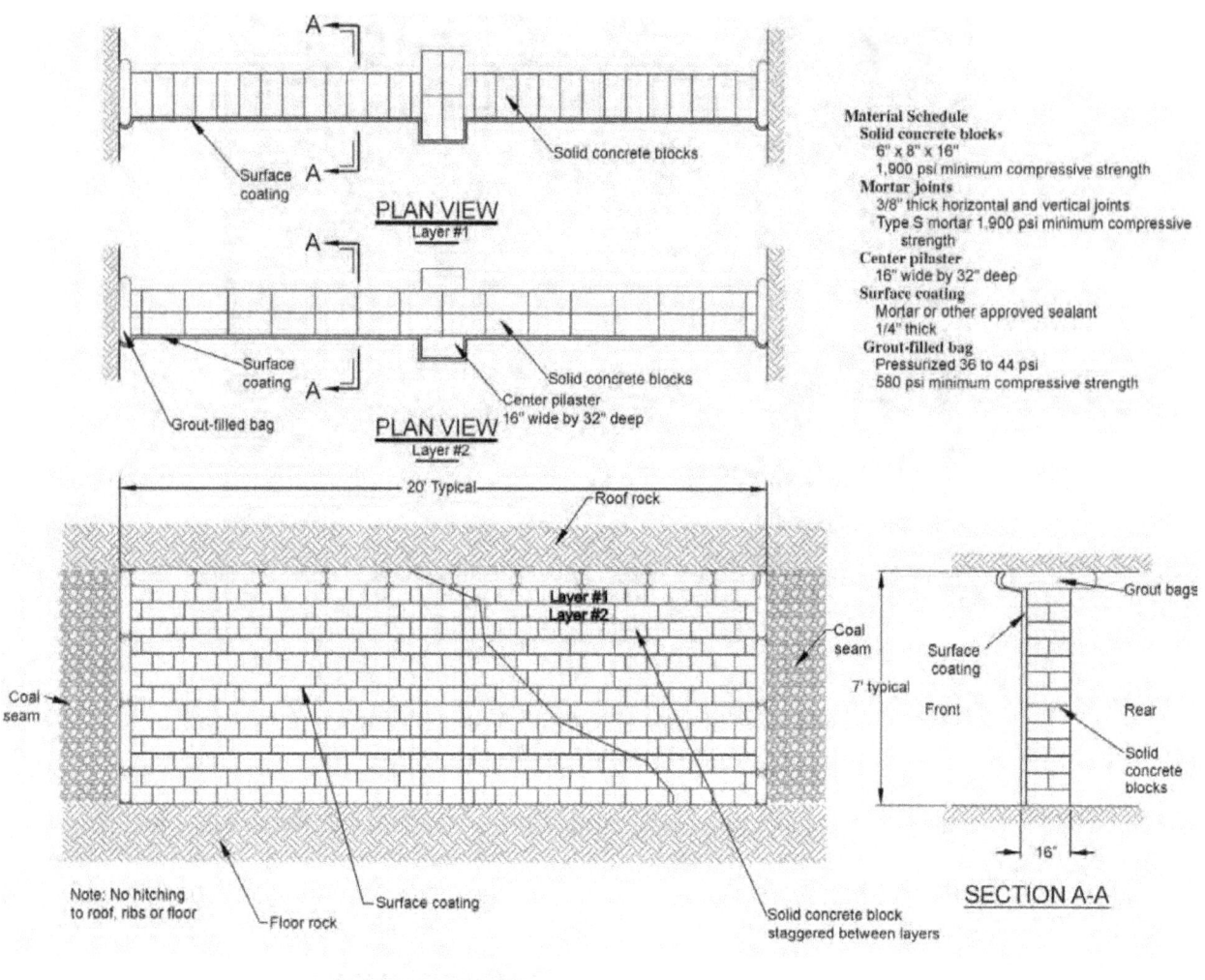

Figure 30.—Front-, plan-, and side-view drawings of Category 3B structure: solid-concrete-block seal with Packsetter Bags.

After all seal material and Packsetter Bags are in position, grout is pumped into the bags in a sequence beginning with the top-right and top-left bags, followed by the bags along the roof line, followed by the bottom-right and bottom-left bags, and concluding with the bags along the rib lines. Each bag is pressurized with grout to 36–44 psi using a compressor-driven or hand pump, and the bags will load the seal horizontally and vertically upon installation. The Packsetter Bag grout is a specially formulated portland cement-based mixture blended and packaged for Strata Mine Services by Quikrete. It develops a uniaxial compressive strength of 580 psi after 28 days.

After the grout bags are filled, polyurethane foam is used to fill any small gaps at the roof and ribs or between Packsetter Bags. Finally, the outby side of the seal is coated with an MSHA-approved general-purpose sealant to minimize leakage through the seal. Figure 31 shows a nearly complete tongue-and-groove, solid-concrete-block seal with Packsetter Bags.

Figure 31.—Tongue-and-groove, solid-concrete-block seal with center pilaster using pressurized Packsetter grout bags in lieu of hitching around the seal perimeter. The worker is coating the seal with an approved sealant.

Category 3C seal: Ventilation stoppings – solid and hollow-core concrete blocks

Ventilation stoppings are sometimes constructed from dry-stacked solid or hollow-core concrete blocks. Numerous such stoppings have been explosion tested in the LLEM. Although this kind of ventilation stopping is not a seal, it is a type of articulated structure. Solid or hollow-core concrete blocks are often part of seal construction, serving either as an integral part of the seal structure or as a form wall for some other seal construction material. The applied loads and measured responses from tests on these ventilation stoppings may provide useful structural information for analyzing and designing other mine seals, so the test results are included here. NIOSH researchers tested eight different structures in this category (Table 1)—four constructed with solid concrete blocks and four with hollow-core concrete blocks. Weiss et al. [2008] provide additional details on constructing the ventilation stoppings in this category along with test results.

Figure 32 shows front-, plan-, and side-view drawings of the typical ventilation stopping constructed with either solid or hollow-core concrete blocks. The hollow-core concrete blocks measure 6 in by 8 in by 16 in and have a nominal uniaxial compressive strength for the block material of 1,900 psi. Recent full-scale wall testing by Barczak and Batchler [2008] indicates a hollow-core concrete block uniaxial compressive strength of less than 1,000 psi. For three of the solid-concrete-block structures, the blocks also measure 6 in by 8 in by 16 in; for the other structure, the blocks measure 8 in by 8 in by 16 in. The nominal uniaxial compressive strength of the solid-concrete-block material was 1,900 psi. However, recent full-scale wall testing by Barczak and Batchler [2008] indicated a solid-concrete-block compressive strength of 1,330–1,780 psi.

Wood wedges are used to tighten each course of blocks along the ribs and on the top course of blocks against the roof (Figures 32–33). Finally, a layer of Quikrete BlocBond sealant is applied to both faces of the stopping (Figure 34).

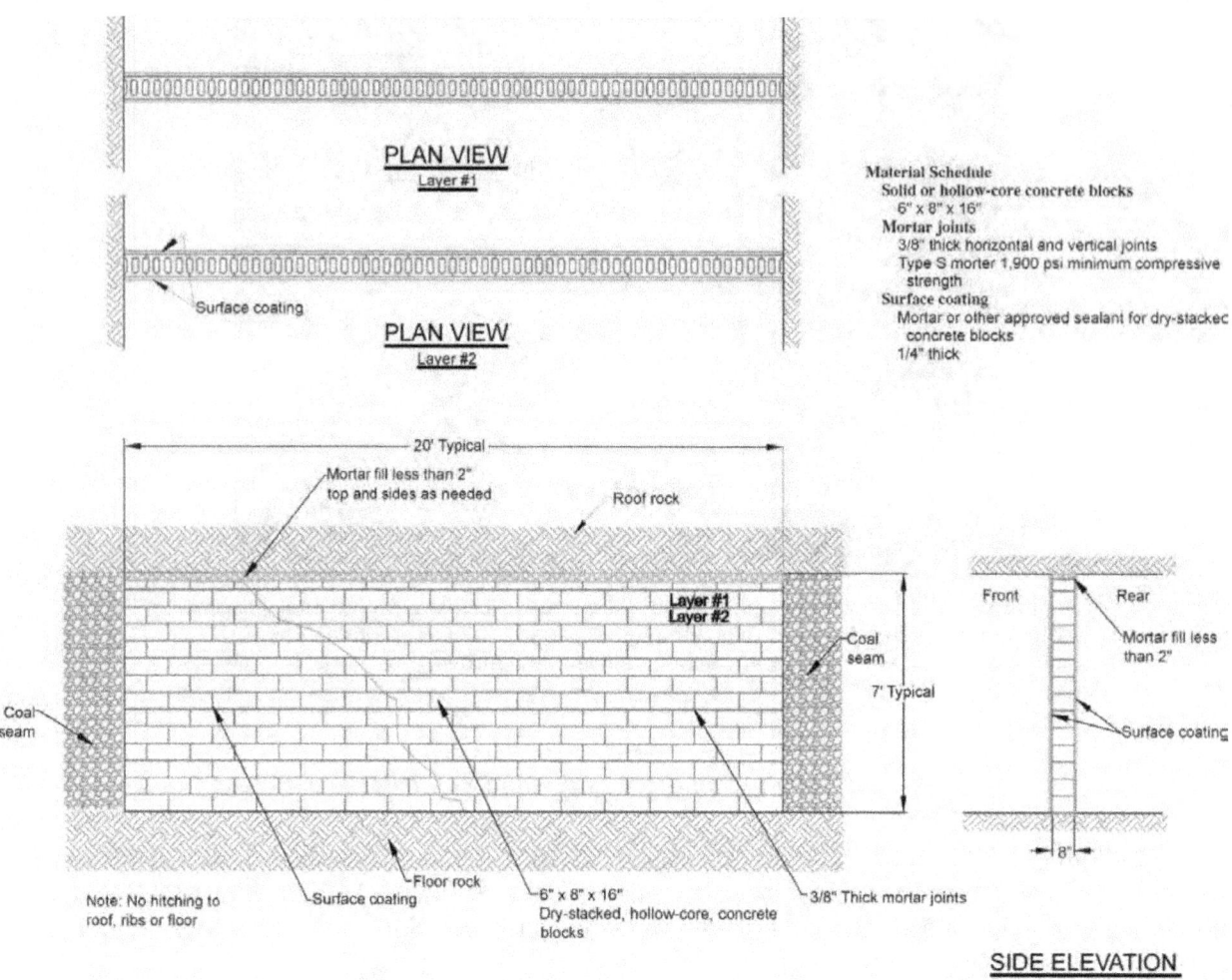

Figure 32.—Front-, plan-, and side-view drawings of Category 3C structure: solid or hollow-core concrete block ventilation stoppings.

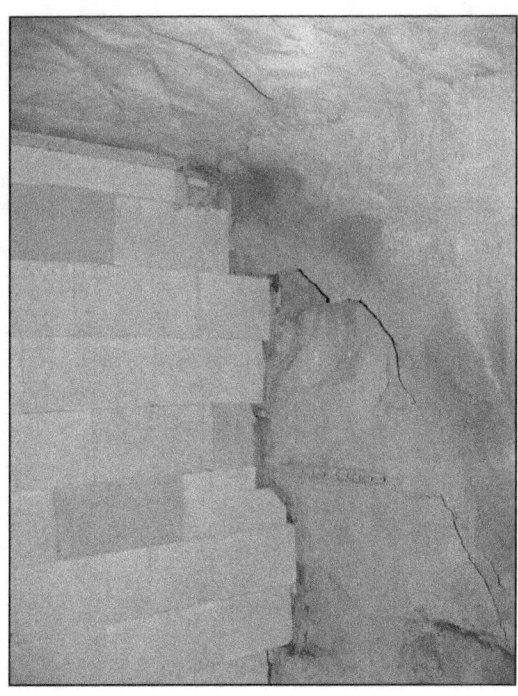

Figure 33.—Dry-stacked concrete block stopping showing wedges used to fit the stopping to ribs and small cut blocks and a wood plank at the top.

Figure 34.—Dry-stacked concrete block stopping showing application of an approved sealant to the surface.

Category 4 Seals: Polymer and Aggregate Materials Without Hitching

The MICON 550 permanent ventilation seal constructed by MICON is the sole representative of the Category 4 seals. This seal is constructed with two dry-stacked hollow-core concrete block stopping walls separated by about 18 in and filled with a mixture of two-component polyurethane foam and crushed limestone aggregate. No hitching of this structure is used except through friction with the surrounding rock. We present data from only one structure in this category (Table 1). Weiss et al. [1996] provide additional details on constructing polymer and aggregate seals in this category along with test results.

Figure 35 shows front-, plan-, and side-view drawings of the MICON 550 seal. The back dry-stacked, hollow-core concrete block wall is constructed first. Upon placement of the last block of each row, a wedge is driven between the block and rib to firmly tighten the blocks in that row. All notches and holes are filled with the largest block fragments possible and wedged in place. The back wall is coated on the outside surface with an MSHA-approved sealant for dry-stacked concrete block walls. Note that MSHA-approved sealants for dry-stacked walls are listed separately from general-purpose sealants.

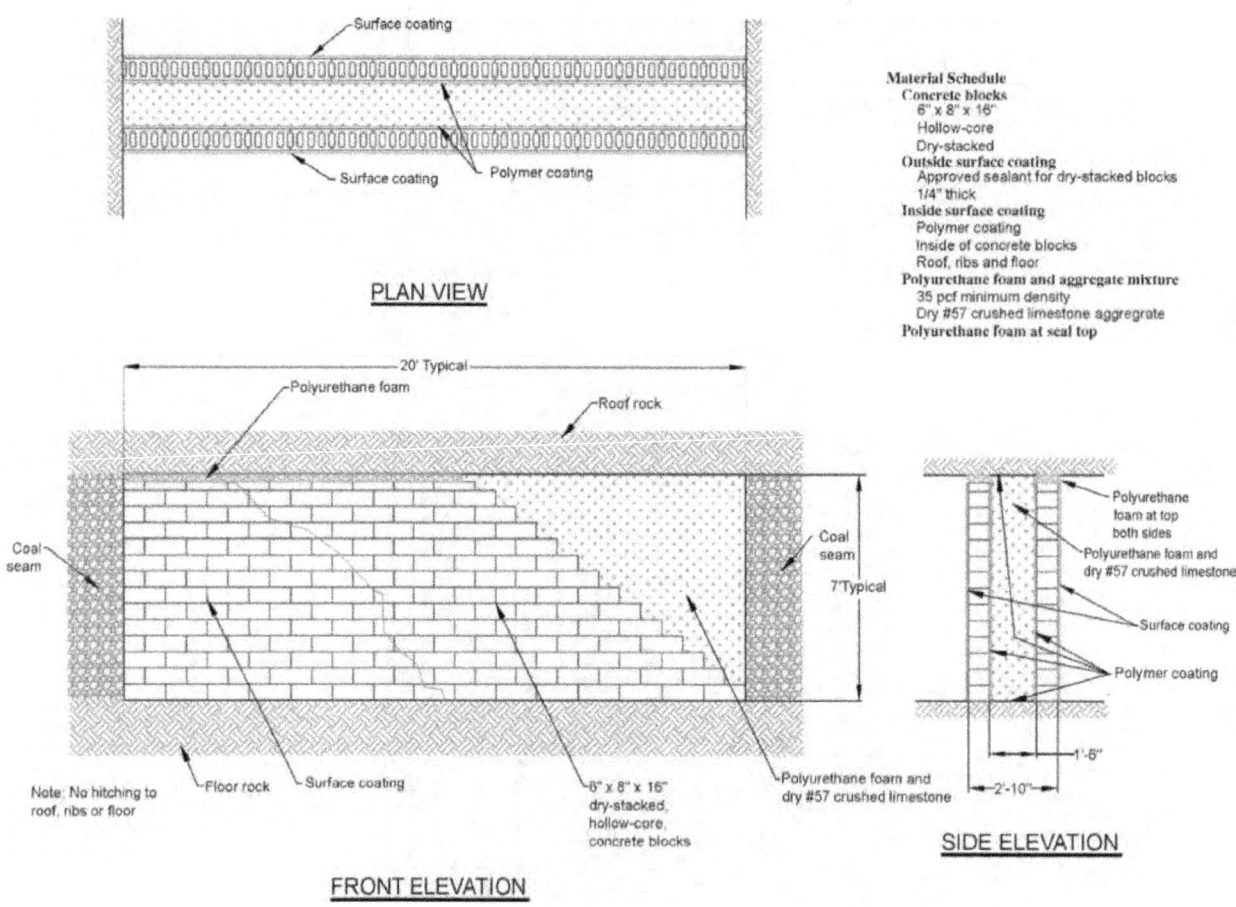

Figure 35.—Front-, plan-, and side-view drawings of Category 4 structure: polymer and aggregate seal.

The front block wall is constructed to a height of 2–3 ft depending on seal height, and the central portion of the front wall is constructed all the way to the roof in a pyramid shape (Figure 36). After one or two top blocks are in place, wedges are driven between the block and roof to hold the wall in place.

Figure 36.—Polymer and aggregate seal from MICON showing the rear dry-stacked, hollow-core concrete block wall, the partially completed front wall, a polymer coating on the inside surface of the form walls, and an approved sealant on the outside surface of the form wall.

With the back wall completed and part of the front wall in place, construction of the inner core can begin. Construction of the inner core continues simultaneously with additional construction of the front wall. The outside of the front wall is also coated with an MSHA-approved sealant for dry-stacked concrete block walls. Note that the concrete blocks must be dry for adequate bonding of the polyurethane foam in the inner core.

The initial step in constructing the inner core is to coat the floor, the inside of the block walls, the roof, and the ribs within the core area with a high-density, 70 lb/ft^3 polymer to prevent moisture from affecting the density of the polyurethane core. This coating on the inner wall is indicated in Figures 35–36. Installation then proceeds as follows:

1. A 4-in-thick layer of dry crushed limestone aggregate, ranging in size from 0.25 to 1 in, is placed as the initial lift.
2. The initial layer of limestone is completely coated with polyurethane foam (Figure 37) having a density of at least 10 lb/ft^3. As the polymer reacts, it expands and rises through with the crushed limestone, then hardens within 5 min.
3. Steps 1 and 2 are repeated to the roof to complete the seal.

The polyurethane foam and aggregate core must achieve a density of at least 35 lb/ft^3 after curing for 24 hr.

The dry-stacked and coated concrete block walls remain in place after seal construction since these walls are an integral part of the structure. If the wall deteriorates, it will require repair. Any exposed polyurethane is coated with an MSHA-approved fire-retardant sealant, both immediately after seal construction and at any later time in the life of the seal.

Figure 37.—Polymer and aggregate seal from MICON showing the rear and front form wall, the polymer coating on the inner surfaces of the form walls, and the polyurethane foam and aggregate mixture filling the inner core of the seal.

Category 5 Seals: Wood-Crib-Block Seals With or Without Hitching

Category 5 seals made mainly from wood crib blocks are intended for use where high convergence is expected. These wood seals crush substantially, but may still retain their ability to resist explosion pressure and not leak air in or out of the sealed area. Two varieties of wood seals are described here: the wood-crib-block seal, which is nailed together and hitched into the surrounding rock; and a variant of this wood seal, which is glued together with construction adhesive and uses pressurized grout bags in lieu of hitching to attach the seal structure firmly to the surrounding rock.

Category 5A seal: Wood-crib-block seal with hitching

Category 5A wood crib seals are used in mines with high convergence where the forces resulting from convergence might crack the standard solid-concrete-block seal, rendering it ineffective. We have no data to report on this structure, but describe it here for completeness. Weiss et al. [1993] provide details on constructing wood-crib-block seals in this category.

Figure 38 shows front-, plan-, and side-view drawings of the typical construction of a modified wood-crib-block seal. Crib blocks are at least 36 in long and measure either 5 in by 5 in or 6 in by 6 in. They are installed horizontally with their length parallel to the ribs. Each crib block is toenailed to the crib block in the lower course using three 4-in-long common nails spaced on 9-in centers, or a 0.5-in-thick layer of rock dust is placed between courses of crib blocks (Figure 39). Voids around the perimeter are wedged tight. In practice, the seal is hitched into the ribs and floor by excavating into solid material at least 6 in. For testing in the LLEM, NIOSH researchers simulate hitching with 6-in by 6-in by 0.5-in-thick steel angle on the floor and ribs (Figure 40). The steel angle is anchored on 18-in centers using 1-in-diam by 9-in-long anchor bolts. Additionally, 5/8-in-thick exterior plywood sheeting is nailed on both sides of the seal using 4-in common nails on 6-in centers. An approved sealant is applied around the perimeter and at the joints in the plywood sheeting on both faces (Figure 40).

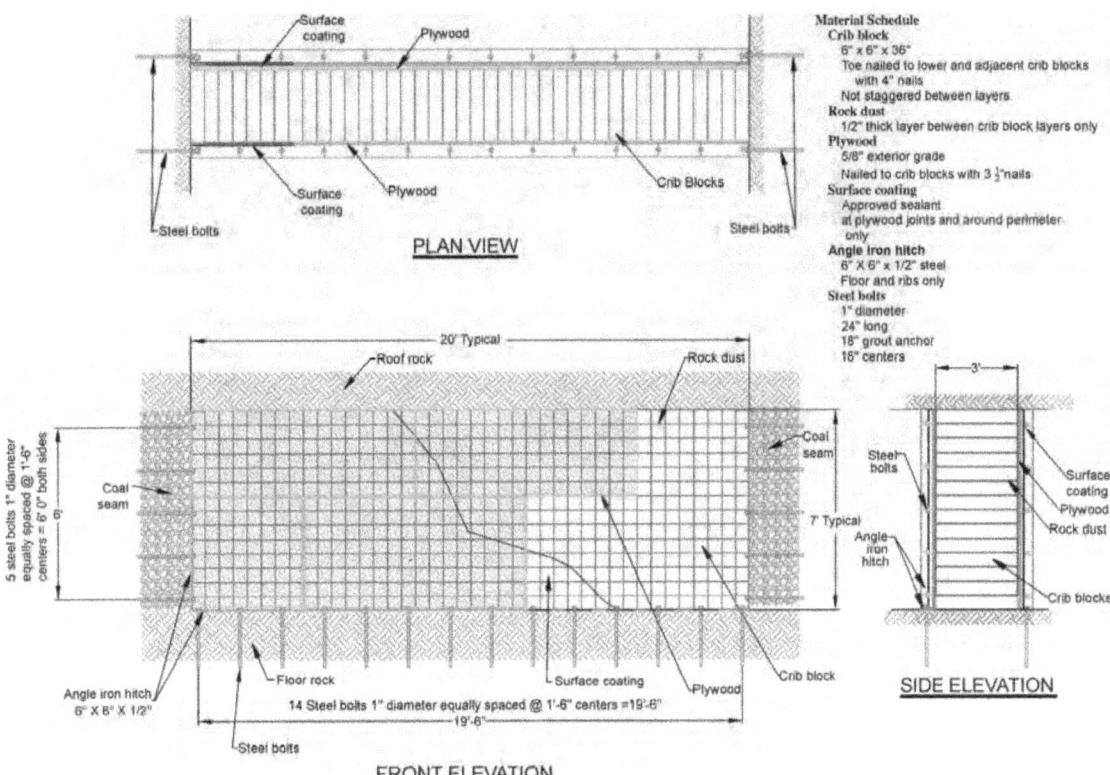

Figure 38.—Front-, plan-, and side-view drawings of Category 5A structure: wood-crib-block seal with plywood facing and simulated hitching.

Figure 39.—Closeup of a hitched wood-crib-block seal showing the layer of rock dust placed between each layer of wood crib blocks.

Figure 40.—Construction of a hitched wood-crib-block seal without plywood facing showing the wood crib blocks separated by a rock dust layer, the angle iron hitching around the ribs and floor only, and the final coating with an approved sealant.

Category 5B seal: Wood-crib-block seal with glue and Packsetter Bags

Category 5B wood crib seals are also used in mines that have high convergence where other seals might fail because of cracking. The seals are constructed similar to the ordinary Category 5A wood-crib-block seals, except that Packsetter Bags are used in place of hitching. We present data from only one structure in this category (Table 1). Sapko et al. [2003] provide details on constructing the wood crib block with grout bag seals in this category.

Figure 41 shows front-, plan-, and side-view drawings of the typical construction of a wood-crib-block seal with Packsetter Bags. The seal consists of 30-in-long wood crib blocks that measure from about 5-in by 5-in to 6-in by 6-in, appropriate glue and applicator, and Packsetter Bags with the necessary grout. Construction of the seal begins by applying a layer of adhesive to the mine floor. The first course of wood crib blocks is then laid across the floor or footer in the adhesive. The length of the crib block is oriented parallel with the ribs. No hitching to the roof, ribs, or floor is necessary.
A Packsetter Bag is positioned under the corners of the first course of crib blocks by at least 6 in. The crib blocks are glued together with FOMO Handi-Stick adhesive and a special applicator (Figure 42). Each crib block requires three rows of adhesive about 0.5 in wide applied to the top and each side of each crib block along its entire length.

In subsequent rows of crib blocks, the vertical joints are staggered (Figures 41 and 43). The wood crib blocks are placed to within 2–5 in of the roof and ribs to allow space for the Packsetter Bags. The bags along the ribs and across the roof overlap adjacent bags by a minimum of 6 in and the front and back face of the seal by at least 3 in. The top-right and top-left corner Packsetter Bags are placed with half the bag down the rib side and half the bag across the roof side of the seal. A spacer Packsetter Bag is used to fill gaps up to 8 in caused by rib and roof undulations. The Packsetter Bag grout is a specially formulated portland cement-based mixture blended and packaged for Strata Products, Inc., by Quikrete and has a compressive strength of 580 psi after 28 days.

After wood crib blocks and Packsetter Bags are in position, grout is pumped into the bags to a pressure of 50 psi in a sequence beginning with the top-right and top-left bags, followed by the bags along the roof line, then the bottom-right and bottom-left bags, and concluding with the bags along the rib lines. This pressurization provides artificial horizontal and vertical loading on the seal upon installation. Polyurethane foam is used to fill any small gaps at the roof, ribs, or between Packsetter Bags. The active mine side of the seal is covered with brattice secured with several pieces of wood nailed to the seal. The brattice overlaps the seal perimeter at the roof, rib, and floor by at least 1 ft. Finally, the active mine side of the seal is coated with an MSHA-approved general-purpose sealant.

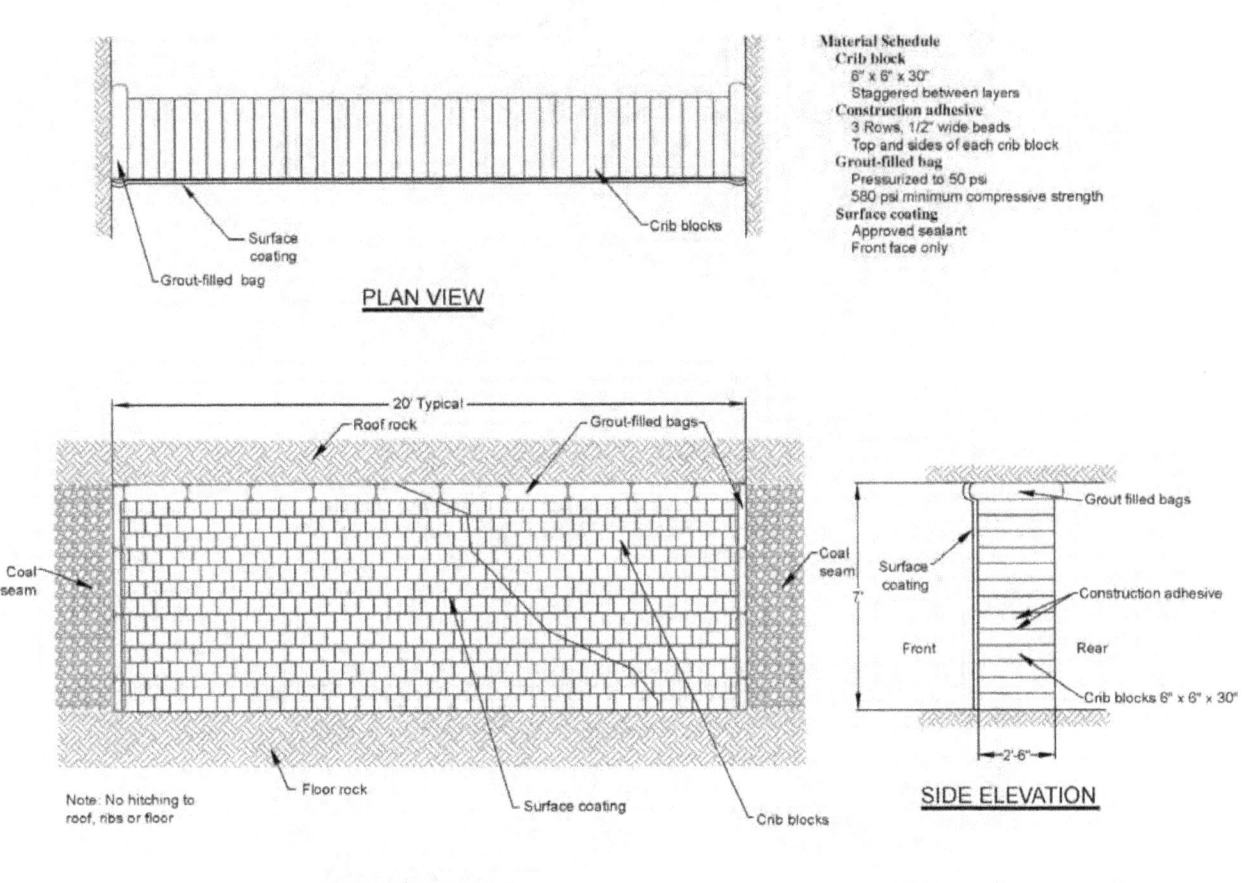

Figure 41.—Front-, plan-, and side-view drawings of Category 5B structure: wood-crib-block seal with Packsetter Bags.

Figure 42.—Construction of a wood-crib-block seal with pressurized Packsetter grout bags in lieu of hitching showing the application of glue between all wood-crib-block surfaces.

Figure 43.—Construction of a wood-crib-block seal with pressurized Packsetter grout bags in lieu of hitching showing the glued wood crib blocks and the grout bags used along the roof and ribs. The active-mine-side surface is later covered with brattice and coated with an approved sealant.

Category 6 Seals: Articulated Structures – Lightweight Blocks With or Without Hitching

Category 6 seals are articulated structures made from lightweight Omega blocks manufactured by Burrell Mining Products International, Inc. The Omega blocks are noncombustible, glass-fiber reinforced blocks measuring 8 in by 16 in by 24 in, weighing about 46 lb and having a compressive strength of 70–110 psi. The blocks can be cut with a handsaw to fit into spaces between seal and rib or roof where full blocks will not fit. Two varieties of seal structures composed of lightweight blocks are reported here: 24-in-thick seals that require hitching and 40-in-thick seals that do not require hitching.

In Omega block seal construction, all horizontal and vertical joints are coated with 0.25-in-thick BlocBond mortar manufactured by Quikrete Company. BlocBond, a mortar mix containing portland cement, fiberglass fiber, and additives, is the only mortar permitted in the construction of Omega Block seals. A layer of BlocBond is applied to the floor for setting the first course of blocks. BlocBond is also placed between the blocks and the coal ribs and between the top blocks and the mine roof. Loose material is brushed off of each block, then blocks are wetted before applying the BlocBond. Finally, all outside faces are coated completely with a 0.25-in-thick layer of the BlocBond mortar.

Category 6A seal: Lightweight blocks 24 in thick with hitching

The Category 6A 20-psi seal designs are made with Omega blocks constructed 24 in thick with a 48-in by 48-in center pilaster for added strength. NIOSH researchers tested two different structures in this category (Table 1). Sapko et al. [2005] and Cashdollar et al. [2007] provide additional details on constructing the lightweight block seals in this category along with test results.

Figure 44 shows front-, plan-, and side-view drawings, and Figure 45 shows the construction of this Omega block seal with hitching. All block surfaces are coated with a 0.25-in-thick layer of BlocBond mortar. Two wood planks are placed across the top of the blocks and wedged against the roof (Figure 46). Wood planks are also placed on top of each pilaster and wedged against the roof. Spaces between the wood planks and wedges are filled with BlocBond mortar. To simulate hitching during the LLEM tests, NIOSH researchers use a 6-in by 6-in by 0.5-in-thick steel angle on the floor and ribs that is anchored on 18-in centers using 1-in-diam by 9-in-long Hilti Kwik Bolts. Figure 47 shows a completed Omega block seal with the angle iron to simulate hitching.

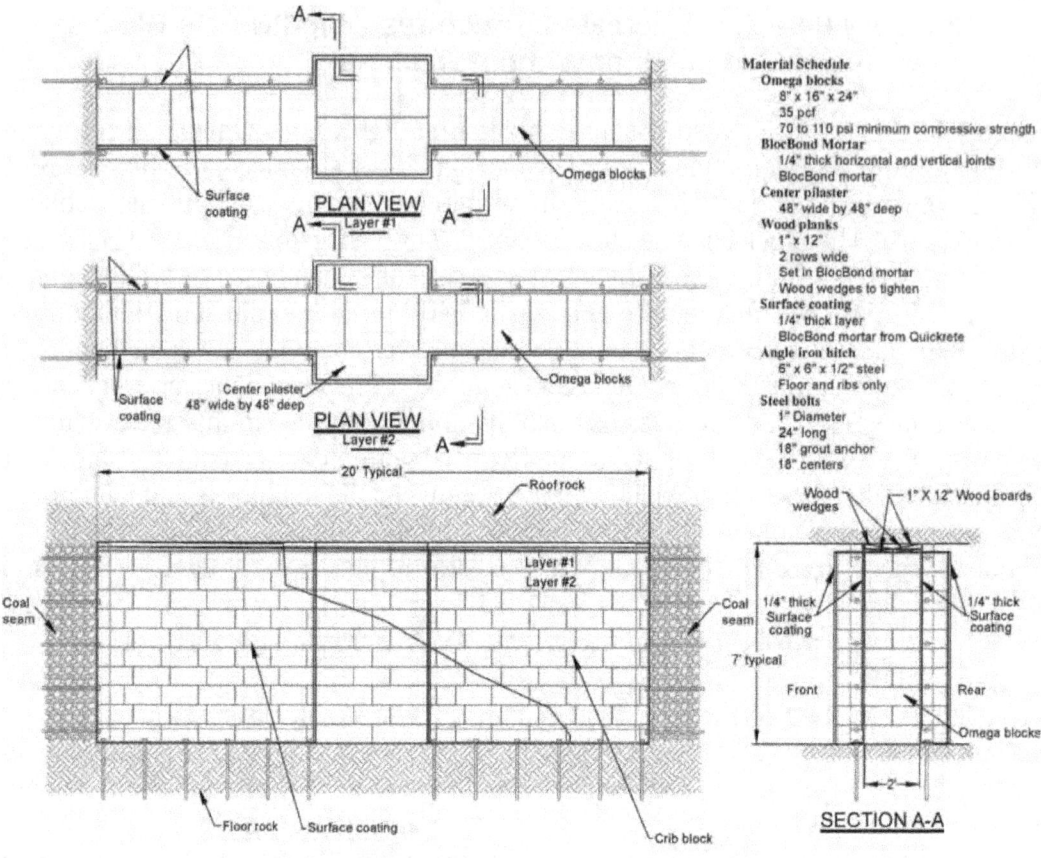

Figure 44.—Front-, plan-, and side-view drawings of Category 6A structure: lightweight block seal, 24 in thick with hitching.

Figure 45.—Construction of a 24-in-thick Omega block seal with hitching and a center pilaster. All surfaces are coated with BlocBond mortar.

Figure 46.—Construction of an Omega block seal showing the wood board and wedges used to secure the seal at the roof.

Figure 47.—Completed 24-in-thick Omega block seal showing the angle iron hitch along the ribs and floor and the outer surface coated with an approved sealant.

Category 6B seal: Lightweight blocks 40 in thick without hitching

The Category 6B 20-psi seal designs are made with Omega blocks constructed 40 in thick and without a center pilaster for added strength or simulated hitching with steel angle at the floor and ribs. NIOSH researchers tested eight different structures in this category (Table 1). Weiss et al. [1993, 1996], Weiss and Harteis [2008], Sapko et al. [2003], and Cashdollar et al. [2007] provide additional details on constructing the lightweight block seals in this category along with test results.

Figure 48 shows front-, plan-, and side-view drawings of the typical Omega block seal without hitching. Figure 49 shows the same seal design under construction. Joints in alternate courses of blocks are staggered. Three rows of 1-in by 12-in wood planks are placed from rib to rib across the top of the seal. One row of planks is placed along the center of the seal; the other two rows are placed with their edges flush with the inby and outby faces of the seal, respectively. Joints in the planking are staggered, and the planking is set in a thin layer of BlocBond. Wedges are driven on 6- to 12-in centers between the planks and roof to compress the planks uniformly against the Omega blocks. BlocBond is used to fill all the gaps between the mine roof and the top block course, including the areas between the rows of wood planks and the gaps between the wooden wedges. Finally, all exposed wood and the outside faces are coated with BlocBond.

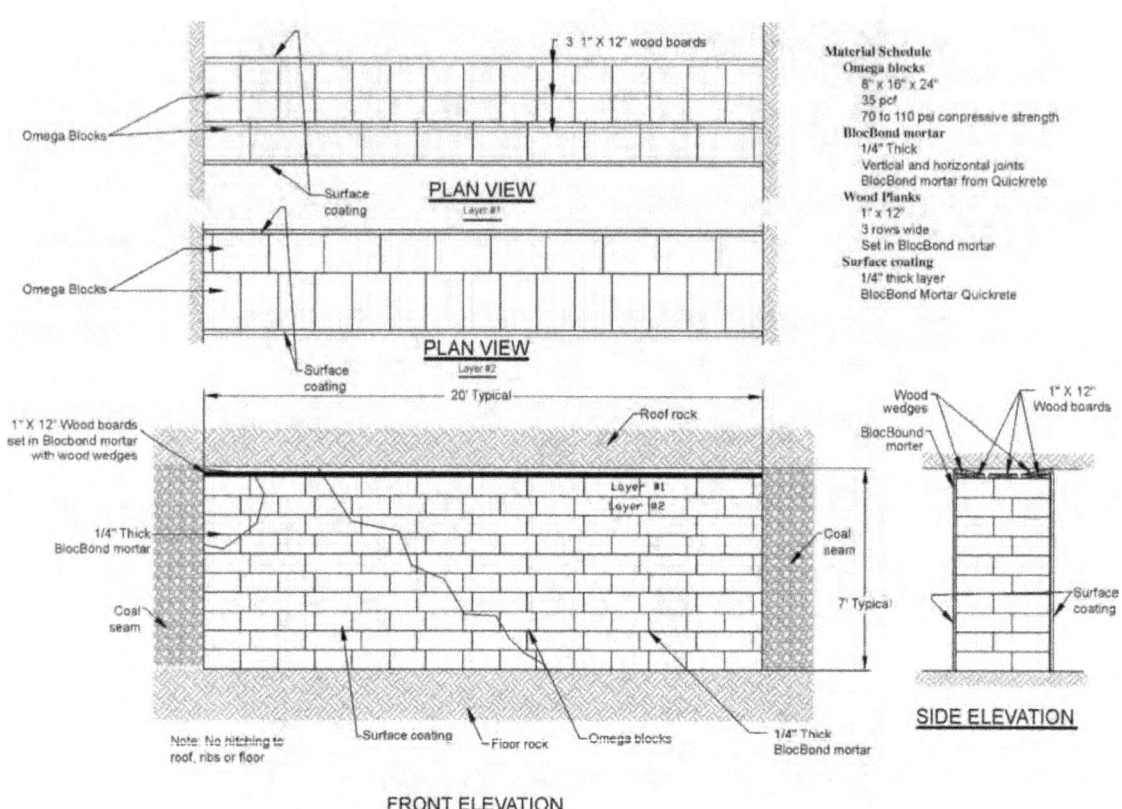

Figure 48.—Front-, plan-, and side-view drawings of Category 6B structure: lightweight block seal, 40 in thick without hitching.

Figure 49.—Construction of a 40-in-thick Omega block seal without hitching and no center pilaster. All surfaces are coated with BlocBond mortar.

STRUCTURAL TESTING DATA FOR 20-psi COAL MINE SEALS

Table 1 provides an overview of the NIOSH testing program on 44 different seal structures and 8 stoppings. As the table indicates, the program subjected many structures to multiple or repeated pressure loads and tested many structures to failure. For some of the structures tested, the only data available are the applied P-t curve and an indicator of whether the structure survived or failed under that applied load. However, for most of the tests, the applied P-t curve and the measured D-t response are reported here.

Tables 5 through 10 provide expanded information for each of the six seal categories, including the structure number for each individual seal structure constructed and tested, the type and number of tests on that structure, the loading condition on that structure (static or reflected, nonuniform or uniform), a reference figure that most closely describes that structure, and a brief description of the structure. Again, the test program includes 52 separate structures (44 seals and 8 stoppings).

Table A-1 in the appendix provides detailed information for each seal structure and each test on that structure. Column 1 in the table gives the seal category and structure, a brief description, the name of the manufacturer or constructor, NIOSH publications related to that particular seal structure, and the name of the Excel file containing the raw data for the P-t and D-t curves associated with that seal. The Excel data files will be posted and available on the NIOSH Mining Web site (http://www.cdc.gov/niosh/mining).

Column 2 in Table A-1 gives the location where the structure was constructed and tested; the exact height, width, and thickness for that particular seal structure; the type of test from previous tables; and the loading condition (also from previous tables). For most of the LLEM explosion tests, the location is the crosscut number between either the A- and B-drifts or B- and C-drifts. Other structures were tested in the small or large hydrostatic test chambers in the LLEM.

For each unique structure tested, column 3 in Table A-1 gives a detailed description of how that seal was constructed. This information comes directly from the LLEM test descriptions. Although the construction details will generally follow the descriptions given in the previous section, each structure tested is unique. For example, the shotcrete thickness in the Category 1A seals will vary from structure to structure, and the steel reinforcement may change somewhat from structure to structure. Those details, if available, are provided in the description in column 3.

Column 4 in Table A-1 gives the LLEM test numbers that subjected the particular structure to an explosion loading. Each explosion test in the LLEM since the inception of the facility in 1983 has been assigned a sequential number. As indicated in Table 1, most of the structures reported herein were subjected to multiple explosion loadings prior to either failure of that structure or cessation of further testing. Note that individual LLEM tests will generally load from three to five different structures during the same test.

Column 5 in Table A-1 indicates the test outcome. Either the structure withstood the explosion without significant damage, was damaged to some degree, or failed catastrophically, foregoing the possibility of any further test loadings.

Column 6 in Table A-1 gives the maximum pressure to which the structure was subjected during the explosion test. The pressure data reported are from the raw test data (nonsmoothed).

Column 7 in Table A-1 gives the maximum displacement recorded, if available. As indicated in Table 1, D-t data from an LVDT exist for most of the 52 structures tested and are reported here. In many of the earlier explosion tests of seal structures, NIOSH researchers did not measure the D-t response of the structure with an instrument. Seal testing and approval at the time were based on a pass or fail explosion test, and there was usually no immediate need for measuring the structural response.

Column 8 in Table A-1 refers to the figure in the appendix that graphically displays the applied P-t curve and the measured D-t response of the structure to each explosion or water pressure test. These plots and the data they contain are valuable for calibrating and verifying computer models for the structural response of seals. Most of these plots had not been previously published, as such structural information was not necessary in the "build and test" approach used for the previous 20-psi seal designs [Gadde et al. 2007].

Data available for certain seal subcategories should enable model calibration and extrapolation to new seal designs and possible development of structure failure criteria for various seal designs. Available data for other seal subcategories are less complete, since NIOSH researchers may have conducted studies in those subcategories prior to 1997 and that information is no longer available. Below are discussions on the structural data available for each seal subcategory.

Category 1A: Insteel 3-D seal – shotcrete with reinforcement.—This data set includes tests on seven different structures where all seven data sets have both the P-t and D-t curves. For structural analysis, the best data sets come from structures 1, 4, and 7, which were constructed similar to the schematic shown in Figure 15. Structures 1 through 6 were subject to a static, nonuniform explosion pressure that swept across the seal face, whereas structure 7 was subject to a static, uniform, water pressure load from a hydrostatic chamber test (Table 5). Test data from structures 4 and 7, plus data from structure 1, which is a thinner version of the same, should provide useful information for developing structural failure criteria for these seal designs and similar structures. Structures 2, 3, 5, and 6 are variations of the seal geometry shown in Figure 15; these data sets may also contribute to model calibration and failure criteria development.

Category 1B: Meshblock seal – shotcrete with reinforcement.—This data set includes tests on six different structures. Unfortunately, good D-t data are not available from any of the tests. Structures 2 and 3 were constructed exactly as shown in Figure 20. Structures 4, 5, and 6 are thinner versions of the previous structures, but are constructed similar to the schematic shown in Figure 20. Structure 1 has some resemblance to the construction shown in Figure 20, but is much thicker and may be more akin to the Category 2 pumpable cementitious material seals. Each structure was loaded repeatedly by a static, nonuniform explosion pressure that swept across the seal face (Table 5). Three of those structures survived and three failed during the last test. This data set should provide a good basis for model calibration and possible development of a failure criterion.

Table 5.—Summary of Category 1 seal structures: concrete with steel reinforcement

Structure No.	Type of test	No. of tests	Loading condition	Reference figure	Structure description
CATEGORY 1A					
Insteel 3-D seal: Concrete or concretelike materials with internal steel reinforcement and anchorage to rock					
1	Explosion	1	Static Nonuniform	Similar to Figure 15	1 Insteel 3-D panel, 1 plane of steel reinforcement, 7-in-thick shotcrete
2	Explosion	1	Static Nonuniform	Similar to Figure 15	1 Insteel 3-D panel, 1 plane of steel reinforcement, 3-in-thick foam core, 3-in-thick shotcrete on both sides
3	Explosion	1	Static Nonuniform	Similar to Figure 15	1 Insteel 3-D panel, 1 plane of steel reinforcement, 3-in-thick foam core, 1.5-in-thick shotcrete on one side, 3-in-thick shotcrete on other side
4	Explosion	1	Static Nonuniform	Identical to Figure 15	2 Insteel 3-D panels, 2 planes of steel reinforcement, 12-in-thick shotcrete
5	Explosion	1	Static Nonuniform	Based on Figure 15	1 Insteel 3-D panel, 1 plane of steel reinforcement, 8-in-thick shotcrete, 48-in-wide pilaster in center
6	Explosion	1	Static Nonuniform	Based on Figure 15	1 Insteel 3-D panel, 1 plane of steel reinforcement, 8-in-thick shotcrete, 48-in-wide pilaster in center
7	Hydrostatic	1	Static Uniform	Identical to Figure 15	2 Insteel 3-D panels, 2 planes of steel reinforcement, 12-in-thick shotcrete
CATEGORY 1B					
Meshblock seal: Concrete or concretelike materials with internal steel reinforcement and anchorage to rock					
1	Explosion	20	Static Nonuniform	Similar to Figure 20	1 plane of steel reinforcement, 4-ft-thick seal with 533-psi compressive strength material
2	Explosion	5	Static Nonuniform	Identical to Figure 20	13 in thick with 4,200-psi compressive strength material
3	Explosion	3	Static Nonuniform	Identical to Figure 20	13 in thick with 4,200-psi compressive strength material
4	Explosion	2	Static Nonuniform	Similar to Figure 20	7 in thick with 4,200-psi compressive strength material
5	Explosion	2	Static Nonuniform	Similar to Figure 20	3 in thick with 4,200-psi compressive strength material
6	Explosion	2	Static Nonuniform	Similar to Figure 20	1.75 in thick with 4,200-psi compressive strength material

Category 2A: Pumpable cementitious material with compressive strength of 200 psi constructed 48 in thick.—This data set includes tests on four different structures, all of which have good P-t and D-t data. Two of the structures were loaded repeatedly, and three were eventually loaded to failure. Construction of these structures is identical to that shown in Figure 23. Structure 1 was subject to a reflected explosion pressure that was uniform across the seal face; structures 2, 3, and 4 were subject to static, uniform pressure developed by methane ignition tests and water pressure tests in a hydrostatic test chamber (Table 6). This data set should provide an excellent basis for model calibration and possible development of a failure criterion.

Category 2B: Pumpable cementitious material with compressive strength of 433 psi constructed 36 in thick.—This data set includes just one structure identical to that shown in Figure 23. The structure was loaded twice by a static, nonuniform explosion pressure that swept across the seal face, but it did not fail. While P-t data are available, D-t data are not. This data set provides a limited basis for model calibration and extrapolation; however, it provides some data on the structural behavior of stronger construction materials.

Category 2C: Pumpable cementitious material with compressive strength of 480–600 psi constructed 24–30 in thick.—This data set includes tests on four different structures, all of which were loaded repeatedly and all of which survived. D-t data are available for just one of these structures. Construction of these structures is identical to that shown in Figure 23. The structures were loaded by a static, nonuniform explosion pressure that swept across the seal face. This data set provides a limited basis for model calibration and extrapolation; however, it does provide additional data on the structural behavior of stronger construction materials.

Table 6.—Summary of Category 2 seal structures: pumpable cementitious materials

Structure No.	Type of test	No. of tests	Loading condition	Reference figure	Structure description
CATEGORY 2A Pumpable cementitious materials: 200 psi and 48 in thick, no steel reinforcement and no hitching					
1	Explosion	2	Reflected Uniform	Similar to Figure 23	48 in thick with 143-psi compressive strength material with retrofit
2	Hydrostatic	1	Static Uniform	Identical to Figure 23	48 in thick with 350-psi compressive strength material
3	Hydrostatic	3	Static Uniform	Identical to Figure 23	48 in thick with 350-psi compressive strength material
4	Hydrostatic	1	Static Uniform	Identical to Figure 23	48 in thick with 350-psi compressive strength material
CATEGORY 2B Pumpable cementitious materials: 433 psi and 36 in thick, no steel reinforcement and no hitching					
1	Explosion	2	Static Nonuniform	Identical to Figure 23	36 in thick with 450- to 500-psi compressive strength material
CATEGORY 2C Pumpable cementitious materials: 480–600 psi and 24–30 in thick, no steel reinforcement and no hitching					
1	Explosion	2	Static Nonuniform	Similar to Figure 23	34 in thick with 600- to 840-psi compressive strength material, 1 plane of steel reinforcement
2	Explosion	2	Static Nonuniform	Identical to Figure 23	30 in thick with 390- to 565-psi compressive strength material
3	Explosion	2	Static Nonuniform	Identical to Figure 23	24 in thick with 600- to 750-psi compressive strength material
4	Explosion	4	Static Nonuniform	Identical to Figure 23	30 in thick with 597-psi compressive strength material

Category 3A: Standard solid-concrete-block seal with hitching.—This data set includes tests on seven distinct structures, five of which were subject to repeated pressure loads and three of which were loaded to failure. Complete P-t data are presented for each test on all seven structures, and D-t data are available for each test on six of the seven structures in this subcategory. Construction of these structures is identical to that shown in Figure 26, except for structure 1, which lacks the center pilaster. The structures were loaded by different methods (Table 7). Structures 1 through 3 were loaded by a static, nonuniform explosion pressure that swept across the seal face. Structures 4 through 7 were loaded by a static, uniform pressure developed by methane ignition tests and water pressure tests in a hydrostatic test chamber. This data set should provide an excellent basis for model calibration and possible development of a failure criterion.

Category 3B: Solid-concrete-block seal with Packsetter Bags and without hitching.—This data set includes tests on three structures where only one was loaded to failure. P-t data are available for all tests; however, no D-t data were recorded. Construction of structure 2 is identical to that shown in Figure 30, whereas structures 1 and 3 are similar. The structures were loaded by a static, nonuniform explosion pressure that swept across the seal face (Table 7). This data set may be combined with data from Subcategory 3A and 3C structures for model calibration and possible development of a failure criterion.

Category 3C: Ventilation stoppings – solid and hollow-core concrete blocks.—This data set contains tests on eight different structures where all were loaded repeatedly and all failed eventually. Complete P-t data are available for all tests on all structures, and complete D-t data are available for all but one of the structures tested. Construction of these structures is identical to that shown in Figure 32. All structures were loaded by a static, nonuniform explosion pressure that swept across the stopping face (Table 7). Although this data set is not for mine seals per se, it has significant value for model calibration and development of failure criteria since concrete blocks, both solid and hollow-core, can be a significant part of many seal designs. These tests provide information on the arching behavior of articulated structures on a rigid foundation [Weiss et al. 2008].

Category 4: Polymer and aggregate materials without hitching.—This data set includes just one test on one structure, which resulted in failure. Both P-t and D-t data are presented. Dolinar et al. [2008] provide additional, new structural data on polymer and aggregate seals that were not incorporated into this report. Construction of this structure is similar to that shown in Figure 35. The structure was loaded by a static, uniform pressure developed by a methane ignition test in a hydrostatic test chamber (Table 8). This single test provides useful information for model calibration and possible development of a failure criterion.

Table 7.—Summary of Category 3 seal structures: articulated structures

Structure No.	Type of test	No. of tests	Loading condition	Reference figure	Structure description
	CATEGORY 3A Articulated structures: Standard solid-concrete-block seal with hitching				
1	Explosion	4	Static Nonuniform	Similar to Figure 26	No center pilaster, but with hitching
2	Explosion	10	Static Nonuniform	Identical to Figure 26	Center pilaster and with hitching
3	Explosion	2	Static Nonuniform	Identical to Figure 26	Center pilaster and with hitching
4	Hydrostatic	5	Static Uniform	Identical to Figure 26	Center pilaster and with hitching
5	Hydrostatic	2	Static Uniform	Identical to Figure 26	Center pilaster and with hitching
6	Hydrostatic	1	Static Uniform	Identical to Figure 26	Center pilaster and with hitching
7	Hydrostatic	1	Static Uniform	Identical to Figure 26	Center pilaster and with hitching
	CATEGORY 3B Articulated structures: Standard solid-concrete-block seal with Packsetter Bags and without hitching				
1	Explosion	2	Static Nonuniform	Similar to Figure 30	Center pilaster and with hitching, mortared joints
2	Explosion	2	Static Nonuniform	Identical to Figure 30	Center pilaster and no hitching, mortared joints
3	Explosion	2	Static Nonuniform	Similar to Figure 30	Center pilaster and no hitching, dry-stacked blocks
	CATEGORY 3C Articulated structures: Ventilation stoppings – solid and hollow-core concrete blocks				
1	Explosion	2	Static Nonuniform	Identical to Figure 32	Dry-stacked, hollow-core concrete blocks
2	Explosion	2	Static Nonuniform	Identical to Figure 32	Dry-stacked, hollow-core concrete blocks
3	Explosion	6	Static Nonuniform	Identical to Figure 32	Dry-stacked, hollow-core concrete blocks
4	Explosion	7	Static Nonuniform	Identical to Figure 32	Dry-stacked, hollow-core concrete blocks
5	Explosion	6	Static Nonuniform	Identical to Figure 32	Dry-stacked, solid concrete blocks
6	Explosion	7	Static Nonuniform	Identical to Figure 32	Dry-stacked, solid concrete blocks
7	Explosion	10	Static Nonuniform	Identical to Figure 32	Dry-stacked, solid concrete blocks
8	Explosion	10	Static Nonuniform	Identical to Figure 32	Dry-stacked, solid concrete blocks

Table 8.—Summary of Category 4 seal structures: polymer and aggregate structures

Structure No.	Type of test	No. of tests	Loading condition	Reference figure	Structure description
	CATEGORY 4 Polymer and aggregate materials without hitching				
1	Hydrostatic	1	Static Uniform	Similar to Figure 35	11-in-thick polyurethane foam and aggregate core

Category 5A: Wood-crib-block seal with hitching.—No structural data are available.

Category 5B: Wood-crib-block seal with glue and Packsetter Bags.—This data set contains test results from one structure that was loaded twice and survived both tests. P-t and D-t data are available for both tests. Construction of this structure is identical to that shown in Figure 41. The structure was loaded by a static, nonuniform explosion pressure that swept across the seal face (Table 9). This single test provides useful information for model calibration and possible development of a failure criterion.

Table 9.—Summary of Category 5 seal structures: wood-crib-block structures

Structure No.	Type of test	No. of tests	Loading condition	Reference figure	Structure description
CATEGORY 5A Wood-crib-block seal with hitching					
None	—	0	—	Similar to Figure 38	—
CATEGORY 5B Wood-crib-block seal with glue and Packsetter Bags					
1	Explosion	2	Static Nonuniform	Identical to Figure 41	30-in-thick wood-crib-block seal with Packsetter Bags

Category 6A: Lightweight blocks, 24 in thick with hitching.—This data set contains results from tests on two different structures. All tests have both P-t and D-t data. Construction of these structures is identical to that shown in Figure 44. One structure was loaded by a static, nonuniform explosion pressure that swept across the seal face, while the other was loaded by a static, uniform pressure developed by a methane ignition test in a hydrostatic test chamber (Table 10). Both tests provide useful information for model calibration and possible development of a failure criterion. This data set is included for completeness, although lightweight blocks may or may not have use in future seal designs.

Category 6B: Lightweight blocks, 40 in thick without hitching.—This data set includes results from tests on eight different structures. Three of the structures failed on the first test, three of the structures failed after repeated tests, and two survived repeated tests. Nearly complete P-t and D-t data are available for all tests on all structures. Construction of these structures is identical or very similar to that shown in Figure 48. As indicated in Table 10, four structures were loaded by a static, nonuniform explosion pressure that swept across the seal face. Three were loaded by a reflected explosion pressure that was uniform across the seal face; and one was loaded by a static, uniform pressure developed by a methane ignition test in a hydrostatic test chamber. All tests provide useful information for model calibration and possible development of a failure criterion; however, lightweight blocks may or may not have utility in future seal designs.

Table 10.—Summary of Category 6 seal structures: lightweight block structures

Structure No.	Type of test	No. of tests	Loading condition	Reference figure	Structure description
CATEGORY 6A Articulated structures: Lightweight blocks, 24 in thick with hitching					
1	Explosion	2	Static Nonuniform	Identical to Figure 44	Properly constructed, 24-in-thick Omega block seal with hitching
2	Hydrostatic	1	Static Uniform	Identical to Figure 44	Properly constructed, 24-in-thick Omega block seal with hitching
CATEGORY 6B Articulated structures: Lightweight blocks, 40 in thick without hitching					
1	Explosion	4	Static Nonuniform	Identical to Figure 48	Properly constructed, 40-in-thick Omega block seal without hitching
2	Explosion	9	Static Nonuniform	Identical to Figure 48	Properly constructed, 40-in-thick Omega block seal without hitching
3	Explosion	2	Static Nonuniform	Similar to Figure 48	40-in-thick Omega block seal without hitching, mortar on horizontal joints but not on vertical joints
4	Explosion	3	Static Nonuniform	Similar to Figure 48	40-in-thick Omega block seal without hitching, mortar on horizontal joints but not on vertical joints
5	Explosion	1	Static Uniform	Identical to Figure 48	Properly constructed, 40-in-thick Omega block seal without hitching
6	Explosion	3	Static Uniform	Similar to Figure 48	40-in-thick Omega block seal without hitching, mortar on horizontal joints but not on vertical joints
7	Explosion	1	Static Uniform	Similar to Figure 48	40-in-thick Omega block seal without hitching, mortar on horizontal joints but not on vertical joints
8	Hydrostatic	1	Static Uniform	Identical to Figure 48	Properly constructed, 40-in-thick Omega block seal without hitching

SUMMARY AND CONCLUSIONS

The seal testing data presented in this report are organized into six broad categories of seal structures:

1. Concretelike materials with steel reinforcement and rebar anchorage to rock
2. Pumpable cementitious materials of varying compressive strengths with no steel reinforcement and no hitching
3. Articulated structures such as solid-concrete-block seals and ventilation stoppings made of solid and hollow-core concrete blocks with or without hitching
4. Polymer and aggregate materials without hitching
5. Wood-crib-block seals with or without hitching
6. Articulated structures such as lightweight blocks with or without hitching

This report provides a general description, including figures for each seal category and subcategory. The specific construction detail for individual structures in each seal category is included in the tables in the appendix. The report also describes the LLEM and the general procedures used to load the test seals. NIOSH researchers used four distinct test procedures that subjected the test seals to different loading conditions:

1. Explosion tests on seals in crosscuts that loaded seals with the static blast wave overpressure that is nonuniform across the face.
2. Explosion tests on seals in C-drift that loaded seals with the reflected blast wave overpressure that is assumed uniform across the face.
3. Hydrostatic chamber tests using water pressure that loaded seals with a static, nearly uniform pressure.
4. Hydrostatic chamber tests using methane ignition pressure that loaded seals with a static overpressure that is assumed uniform across the face.

The report discusses and analyzes the loading conditions and boundary conditions for the seal tests. The foundation conditions for the seal tests in the LLEM as reported here do not represent typical conditions found in underground coal mines. The rocks in the LLEM have much greater stiffness and strength than typical rocks found in underground coal mines, where the roof and floor rock and the coal ribs may have lower stiffness and strength.

We report and analyze the sensors and data acquisition systems used during the tests to collect the pressure loading data and the displacement response of the test seals. We report the response time, time constant, and frequency response for various pressure and displacement transducers used throughout the tests. With regard to the P-t data, we conclude that the recorded data accurately reflect the actual pressure developed during the tests. Similarly, the recorded displacement data accurately capture the initial displacement response of the test seals. However, the displacement data are only reliable up to the initial peak displacement and not beyond that point in time due to the experimental method used to link the structure to the displacement transducer.

The structural data reported herein were collected at a rate of 1,500 samples per second. The P-t data could contain higher-frequency components that are not recorded by the data acquisition system. We conducted structural analyses and demonstrated that the P-t data collected at 1,500 samples per second are adequate for structural analysis of seals under the conditions of the explosion tests conducted in the LLEM.

This report presents test data on a total of 52 different structures in the above 6 categories. The appendix presents the P-t curve and the D-t curve for 100 separate tests on these structures. In addition, the appendix provides the P-t curve but not the D-t curve for 66 other tests on the remaining structures. Table A-1 in the appendix indicates whether those structures survived or failed during the test.

The structural data assembled in the appendix will facilitate future structural analysis in each of the various categories of seal structures. Certain data sets may lead to development of structure failure criteria for certain seal designs. Calibrating and validating structural models of seal behavior from 20-psi explosion tests should enable reliable extrapolation of those structural models to the analysis of 50- and 120-psi seal designs intended to meet the MSHA final rule for sealing of abandoned areas.

REFERENCES

73 Fed. Reg. 21182 [2008]. Mine Safety and Health Administration, 30 CFR part 75: sealing of abandoned areas; final rule.

Barczak TM, Batchler TJ [2008]. Comparison of the transverse load capacities of various block ventilation stoppings under arch loading conditions. In: Wallace KG Jr., ed. Proceedings of the 12th U.S./North American Mine Ventilation Symposium (Reno, NV, June 9–11, 2008). Reno, NV: University of Nevada, pp. 225–231.

Cashdollar KL, Weiss ES, Harteis SP, Sapko MJ [2007]. Experimental study of the effect of LLEM explosions on various seals and other structures and objects. Appendix X in: Gates RA, Phillips RL, Urosek JE, Stephan CR, Stoltz RT, Swentosky DJ, Harris GW, O'Donnell JR Jr., Dresch RA. Report of investigation: fatal underground coal mine explosion, January 2, 2006, Sago mine, Wolf Run Mining Company, Tallmansville, Upshur County, West Virginia, ID No. 46-08791. Arlington, VA: U.S. Department of Labor, Mine Safety and Health Administration.

CFR. Code of federal regulations. Washington, DC: U.S. Government Printing Office, Office of the Federal Register.

Dolinar DR [2008]. Personal communication.

Dolinar DR, Sapko MJ, Harteis SP [2008]. Performance of a polyurethane core seal tested in a hydrostatic chamber. Pittsburgh, PA: U.S. Department of Health and Human Services, Centers for Disease Control and Prevention, National Institute for Occupational Safety and Health, DHHS (NIOSH) Publication No. 2008–129, RI 9674.

Esterhuizen GS [2008]. Personal communication.

Gadde MM, Beerbower D, Rusnak JA, Honse J, Worley P [2007]. Post PIB P06-16: how to design alternative mine seals? SME preprint 07-130. Littleton, CO: Society for Mining, Metallurgy, and Exploration, Inc.

Glasstone S, Dolan PJ, eds. [1977]. The effects of nuclear weapons. 3rd ed. U.S. Department of Defense and the Energy Research and Development Administration.

Greninger NB, Weiss ES, Luzik SJ, Stephan CR [1989]. Performance evaluation of solid cementitious seals. Pittsburgh, PA: U.S. Department of the Interior, Bureau of Mines, Pittsburgh Research Center, Internal Report 4809. Unpublished.

Greninger NB, Weiss ES, Luzik SJ, Stephan CR [1991]. Evaluation of solid-block and cementitious foam seals. Pittsburgh, PA: U.S. Department of the Interior, Bureau of Mines, RI 9382. NTIS No. PB 92-152115.

Kinney GF [1962]. Explosive shocks in air. New York: Macmillan.

Landau LD, Lifshitz EM [1987]. Fluid mechanics. Course of theoretical physics, vol. 6. 2nd ed. Oxford, U.K.: Butterwork-Heinemann.

Mattes RH, Bacho A, Wade LV [1983]. Lake Lynn Laboratory: construction, physical description, and capability. Pittsburgh, PA: U.S. Department of the Interior, Bureau of Mines, IC 8911. NTIS No. PB 83-197103.

Sapko MJ, Weiss ES, Trackemas J, Stephan CR [2003]. Designs for rapid in situ sealing. SME preprint 03-010. Littleton, CO: Society for Mining, Metallurgy, and Exploration, Inc.

Sapko MJ, Weiss ES, Harteis SP [2005]. Methods for evaluating explosion-resistant ventilation structures. In: Gillies ADS, ed. Proceedings of the Eighth International Mine Ventilation Congress (Brisbane, Queensland, Australia, July 6–8, 2005). Carlton, Victoria, Australia: Australasian Institute of Mining and Metallurgy, pp. 211–219.

Slawson TR [1995]. Wall response to airblast loads: the wall analysis code (WAC). Prepared for the U.S. Army Energy Research and Development Center, Vicksburg, MS.

Triebsch G, Sapko MJ [1990]. Lake Lynn Laboratory: a state-of-the-art mining research laboratory. In: Proceedings of the International Symposium on Unique Underground Structures. Vol. 2. Golden, CO: Colorado School of Mines, pp. 75-1 to 75-21.

Weiss ES, Harteis SP [2008]. Strengthening existing 20-psi mine ventilation seals with carbon fiber-reinforced polymer reinforcement. Pittsburgh, PA: U.S. Department of Health and Human Services, Centers for Disease Control and Prevention, National Institute for Occupational Safety and Health, DHHS (NIOSH) Publication No. 2008–106, RI 9673.

Weiss ES, Greninger NB, Stephan CR, Lipscomb JR [1993]. Strength characteristics and air-leakage determinations for alternative mine seal designs. Pittsburgh, PA: U.S. Department of the Interior, Bureau of Mines, RI 9477. NTIS No. PB94-111275.

Weiss ES, Slivensky WA, Schultz MJ, Stephan CR. Jackson KW [1996]. Evaluation of polymer construction material and water trap designs for underground coal mine seals. Pittsburgh, PA: U.S. Department of Energy, RI 9634. NTIS No. PB96-123392.

Weiss ES, Cashdollar KL, Mutton IVS, Kohli DR, Slivensky WA [1999]. Evaluation of reinforced cementitious seals. Pittsburgh, PA: U.S. Department of Health and Human Services, Centers for Disease Control and Prevention, National Institute for Occupational Safety and Health, DHHS (NIOSH) Publication No. 99-136, RI 9647.

Weiss ES, Cashdollar KL, Sapko MJ [2002]. Evaluation of explosion-resistant seals, stoppings, and overcast for ventilation control in underground coal mining. Pittsburgh, PA: U.S. Department of Health and Human Services, Centers for Disease Control and Prevention, National Institute for Occupational Safety and Health, DHHS (NIOSH) Publication No. 2003-104, RI 9659.

Weiss ES, Cashdollar KL, Harteis SP, Shemon GJ, Beiter DA, Urosek JE [2008]. Explosion effects on mine ventilation stoppings. Pittsburgh, PA: U.S. Department of Health and Human Services, Centers for Disease Control and Prevention, National Institute for Occupational Safety and Health, DHHS (NIOSH) Publication No. 2009–102, RI 9676.

Zucrow MJ, Hoffman JD [1985]. Gas dynamics. Vol. 2: Multidimensional flow. Malabar, FL: Krieger Publishing Co.

APPENDIX.—DETAILED SUMMARY OF SEAL STRUCTURE TESTS AND TEST DATA

Table A-1.—Detailed summary of seal structure tests

Seal category – Structure No. Description Manufacturer NIOSH reference Excel data file name[1]	Seal test location, dimensions, test type, and loading conditions	Seal description from LLEM test files	LLEM test No.	Test outcome	Maximum pressure data (psi)	Maximum displacement data (in)	Appendix figure No.
CATEGORY 1A SEALS							
Concrete or concretelike materials with internal steel reinforcement and anchorage to rock Insteel 3-D seal: Shotcrete with reinforcement, 7 structures tested							
Category 1A: Structure 1 Shotcrete with reinforcement Precision Mine Repair, Inc. Insteel 3-D Not previously reported by NIOSH Cat1A_Struct#1_1Test.xls	A-B drifts, X-1 19.75 ft wide by 6.67 ft high by 7 in thick Explosion test Quasi-static Nonuniform	1 rear Insteel 3-D panel with Stayform backing and 1 plane of reinforcement bar anchorage similar to Figure 15. The framework, which extends across the entire width of the crosscut, is anchored in place by rebar partially imbedded into the roof, ribs, and floor with the exposed sections of the rebar tie wired to the rebar contained within the framework with 3 on each rib and about 10 each on the roof and floor. Shotcrete (203 bags of Pak Mix Pro Line concrete mix or 15,631 lb dry mix) is then applied to the entire structure from the active (B-drift) side to a total thickness of about 7 in. Nearly fully cured in 24 hr. Minimum compressive strength of shotcrete is 2,500 psi.	419	No damage	45.8	1.0	A-1
Category 1A: Structure 2 Shotcrete with reinforcement Precision Mine Repair, Inc. Insteel 3-D Not previously reported by NIOSH Cat1A_Struct#2_1Test.xls	A-B drifts, X-2 21 ft wide by 6.75 ft high by 9 in thick Explosion test Quasi-static Nonuniform	1 rear Insteel 3-D panel without Stayform backing and 1 plane of reinforcement bar anchorage similar to Figure 15. The framework contains, between the wire mesh outside panels and rebar, a 3-in foam insert. A temporary opening (large enough to allow for the passage of a person) was installed through this seal to allow for the shotcrete application on the sealed (A-drift) side of the seal. 3 in of shotcrete was applied to both sides. A total of 166 bags of Pak Mix Pro Line concrete mix or 12,782 lb of dry mix used for seal. A metal covering was installed within the opening, reinforced with rebar, and completely filled with shotcrete. Nearly fully cured in 24 hr. Minimum compressive strength of shotcrete is 2,500 psi.	419	No damage	33.8	3.2	A-2

[1]Excel data files will be posted and available on the NIOSH Mining Web site (http://www.cdc.gov/niosh/mining).

Table A-1.—Detailed summary of seal structure tests—Continued

Seal category – Structure No. Description Manufacturer NIOSH reference Excel data file name	Seal test location, dimensions, test type, and loading conditions	Seal description from LLEM test files	LLEM test No.	Test outcome	Maximum pressure data (psi)	Maximum displacement data (in)	Appendix figure No.
Category 1A: Structure 3 Shotcrete with reinforcement Precision Mine Repair, Inc. Insteel 3-D Not previously reported by NIOSH Cat1A_Struct#3_1Test.xls	A-B drifts, X-3 About 19 ft wide by 7 ft high by 7.5 in thick Explosion test Quasi-static Nonuniform	1 rear Insteel 3-D panel without Stayform backing and 1 plane of reinforcement bar anchorage similar to Figure 15. The framework contains, between the wire mesh outside panels and rebar, a 3-in foam insert. 1.5 in of shotcrete (Pak Mix Pro Line concrete mix) was applied to the A-drift side. 3 in of shotcrete was applied to the B-drift side. Nearly fully cured in 24 hr. Minimum compressive strength of shotcrete is 2,500 psi.	419	No damage	44.5	3.2	A-3
Category 1A: Structure 4 Shotcrete with reinforcement Precision Mine Repair, Inc. Insteel 3-D Not previously reported by NIOSH Cat1A_Struct#4_1Test.xls	A-B drifts, X-1 18.67 ft wide by 7.5 ft high by 11.5 in thick Explosion test Quasi-static Nonuniform	1 rear Insteel 3-D panel with Stayform backing, 1 plane of reinforcement bar anchorage without Stayform backing, and another plane of reinforcement bar anchorage identical to Figure 15. The framework, which extends across the entire width of the crosscut, was anchored in place by #8 rebar partially imbedded into the roof, ribs, and floor with the exposed sections of the rebar tie wired to the rebar contained within the framework. Shotcrete (362 bags of Pak Mix Pro Line concrete mix or 25,340 lb of dry mix) was then applied to the entire structure from the active (B-drift) side to a total thickness of about 11.5 in. Nearly fully cured in 24 hr. Minimum compressive strength of shotcrete is 2,500 psi.	420	No damage	57.5	0.08	A-4

Table A-1.—Detailed summary of seal structure tests—Continued

Seal category – Structure No. Description Manufacturer NIOSH reference Excel data file name	Seal test location, dimensions, test type, and loading conditions	Seal description from LLEM test files	LLEM test No.	Test outcome	Maximum pressure data (psi)	Maximum displacement data (in)	Appendix figure No.
Category 1A: Structure 5 Shotcrete with reinforcement Precision Mine Repair, Inc. Insteel 3-D Not previously reported by NIOSH Cat1A_Struct#5_1Test.xls	A-B drifts, X-2 About 19 ft wide by 7 ft high by 8 in thick Explosion test Quasi-static Nonuniform	1 rear Insteel 3-D panel with Stayform backing and 1 plane of reinforcement bar anchorage similar to Figure 15. This framework contained a 48-in-wide center pilaster on both sides. A hinged door, the entire height of the seal and about 2 ft wide, was designed on the inby rib end of the seal. This door was then secured with rebar prior to the shotcrete operations. Shotcrete thickness is approximately 8 in. Minimum compressive strength of shotcrete is 2,500 psi.	420	No damage	45.9	3.2	A-5
Category 1A: Structure 6 Shotcrete with reinforcement Precision Mine Repair, Inc. Insteel 3-D Not previously reported by NIOSH Cat1A_Struct#6_1Test.xls	A-B drifts, X-3 18.25 ft wide by 7.4 ft high by about 6 in thick Explosion test Quasi-static Nonuniform	1 rear Insteel 3-D panel with Stayform backing and 1 plane of reinforcement bar anchorage similar to Figure 15. No door through the seal. Diagonal stiffener units on the A-drift side of the framework between the center pilaster and each rib. 255 bags of Pak Mix Pro cement mix or 20,400 lb of dry mix was used. Minimum compressive strength of shotcrete is 2,500 psi. 11.5 in thick at diagonal pilasters and 17.5 in thick at the 48-in-wide center pilaster.	420	No damage	45.1	NA	A-6
Category 1A: Structure 7 Shotcrete with reinforcement Precision Mine Repair, Inc. Insteel 3-D Sapko et al. [2005] Cat1A_Struct#7_1Test.xls	Small hydrostatic test chamber 21 ft wide by 8.5 ft high 11 in thick Water test Quasi-static Uniform	1 rear Insteel 3-D panel with Stayform backing, 1 plane of reinforcement bar anchorage without Stayform backing, and another plane of reinforcement bar anchorage identical to Figure 15. The framework, which extends across the entire width of the crosscut, was anchored in place by #8 rebar partially imbedded into the roof, ribs, and floor with the exposed sections of the rebar tie wired to the rebar contained within the framework. Shotcrete was then applied to the entire structure from the active side to a total thickness of about 11 in. Nearly fully cured in 24 hr. Minimum compressive strength of shotcrete is 2,500 psi.	PR-1	No damage	27	0.22	A-7

Table A-1.—Detailed summary of seal structure tests—Continued

Seal category – Structure No. Description Manufacturer NIOSH reference Excel data file name	Seal test location, dimensions, test type, and loading conditions	Seal description from LLEM test files	LLEM test No.	Test outcome	Maximum pressure data (psi)	Maximum displacement data (in)	Appendix figure No.
		CATEGORY 1B SEALS					
		Concrete or concretelike materials with internal steel reinforcement and anchorage to rock Meshblock seal: Shotcrete with reinforcement, 6 structures tested					
Category 1B: Structure 1 Pumped concretelike materials with reinforcement R. G. Johnson Co., Inc. Aquablend plug seal between Gunmesh stoppings Weiss et al. [1999] Cat1B_Struct#1_16Test.xls	B-C drifts, X-1 17.8 ft wide by 6.4 ft high by 4 ft thick Explosion tests Quasi-static Nonuniform	4-ft-thick seal similar in concept to Figure 20 with 2 reinforced Gunmesh walls sprayed with approximately 0.5-in-thick coating of MB-500 (4,200-psi compressive strength material) and 1 plane of anchored steel reinforcement bar. Aquablend injected between form walls (553-psi compressive strength material). Bolts embedded 24 in into roof/floor/ribs with 24 in extending into entry. Vertical tie-in bolts between roof and floor bolts.	347	No damage	27.0	NA	A-8
			348	No damage	52.5	NA	
			349	No damage	87.7	NA	
			350	No damage	4	NA	
			351	No damage	0.5	NA	A-9
			352, 353	No damage	—	NA	
			354	No damage	35.7	NA	
			355	No damage	25.3	NA	
			356, 357	No damage	—	NA	
			358	No damage	4.2	NA	
			359	No damage	35.9	NA	A-10
			360	No damage	97.4	NA	
			361	No damage	0.5	NA	
			362	No damage	6.2	NA	
			363	No damage	10.0	NA	A-11
			364	No damage	39.1	NA	
			365	No damage	39.6	NA	
			366	No damage	25.7	NA	
Category 1B: Structure 2 Pumped concretelike materials with reinforcement R. G. Johnson Co., Inc. Meshblock Weiss et al. [1999] Cat1B_Struct#2_5Test.xls	B-C drifts, X-2 18.9 ft wide by 7.4 ft high by 13 in thick Explosion tests Quasi-static Nonuniform	12-in-thick seal plus 1-in-thick coating identical to Figure 20. Stacked Meshblock baskets secured with bolts anchored 24 in into roof/floor/ribs. Sprayed MB-500 material (design 4,200 psi) into Meshblock baskets. 5,460-psi compressive strength after 7-day cure and 6,610 psi after 28-day cure. Seal tested on days 12, 19, 27, 40, and 41.	347	No damage	36.1	NA	A-12
			348	No damage	60.0	NA	
			349	No damage	78.7	NA	
			350	No damage	4	NA	A-13
			351	Destroyed	6	NA	

Table A-1.—Detailed summary of seal structure tests—Continued

Seal category – Structure No. Description Manufacturer NIOSH reference Excel data file name	Seal test location, dimensions, test type, and loading conditions	Seal description from LLEM test files	LLEM test No.	Test outcome	Maximum pressure data (psi)	Maximum displacement data (in)	Appendix figure No.
Category 1B: Structure 3 Pumped concretelike materials with reinforcement R. G. Johnson Co., Inc. Meshblock Weiss et al. [1999] Cat1B_Struct#3_3Test.xls	B-C drifts, X-3 19.1 ft wide by 9 ft high by 13 in thick Explosion tests Quasi-static Nonuniform	12-in-thick seal plus 1-in-thick coating identical to Figure 20 in an enlarged section (9-ft-high entry). Stacked Meshblock baskets secured with bolts anchored 24 in into roof/floor/ribs. Sprayed MB-500 material (design 4,200 psi) into Meshblock baskets. Designed as a 24-hr seal. 5,930-psi compressive strength of material after 1 day. Seal tested 27.5 hr after installation and on days 8 and 16.	347 348 349	No damage No damage Destroyed	19.2 74.1 109.1	NA NA NA	A-14
Category 1B: Structure 4 Pumped concretelike materials with reinforcement R. G. Johnson Co., Inc. Meshblock Weiss et al. [1999] Cat1B_Struct#4_2Test.xls	B-C drifts, X-4 19.6 ft wide by 7.4 ft high by 7 in thick Explosion tests Quasi-static Nonuniform	6-in-thick seal plus 1-in-thick coating similar to Figure 20, but not as thick. Stacked Meshblock baskets secured with bolts anchored 24 in into roof/floor/ribs. Sprayed MB-500 material (4,200 psi) into Meshblock baskets. 6,010-psi compressive strength after 7-day cure and 8,778 psi after 28-day cure. Seal tested on days 11 and 18.	347 348	No damage Destroyed	25.0 76.1	NA NA	A-15
Category 1B: Structure 5 Pumped concretelike materials with reinforcement R. G. Johnson Co., Inc. Gunmesh Weiss et al. [1999] Cat1B_Struct#5_2Test.xls	B-C drifts, X-5 19 ft wide by 7.3 ft high by 3 in thick Explosion tests Quasi-static Nonuniform	3-in-thick stopping similar to Figure 20, but not as thick. Stacked Meshblock baskets secured with bolts anchored 24 in into roof/floor/ribs. Sprayed with Tecrete to 3-in thickness. 5,417-psi compressive strength after 7-day cure and 7,330 psi after 28-day cure. Stopping tested on days 13 and 20.	347 348	No damage Destroyed	18.7 59.4	NA NA	A-16

Table A-1.—Detailed summary of seal structure tests—Continued

Seal category – Structure No. Description Manufacturer NIOSH reference Excel data file name	Seal test location, dimensions, test type, and loading conditions	Seal description from LLEM test files	LLEM test No.	Test outcome	Maximum pressure data (psi)	Maximum displacement data (in)	Appendix figure No.
Category 1B: Structure 6 Pumped concretelike materials with reinforcement R. G. Johnson Co., Inc. Gunmesh Weiss et al. [1999] Cat1B_Struct#6_2Test.xls	B-C drifts, X-3 19.3 ft wide by 6.9 ft high by 1.75 in thick Explosion tests Quasi-static Nonuniform	1.75-in-thick stopping similar to Figure 20, but not as thick. Constructed with reinforced Gunmesh screens and sprayed with MB-500 (4,200 psi) and Tecrete. Anchored with bolts embedded 24 in into roof/floor/ribs with 24 in extending into entry. Vertical tie-in reinforcing bars between roof and floor bolts. No samples taken for compressive strength testing. Stopping tested on days 7 and 8.	350 351	No damage Destroyed	3 6	NA NA	A-17
CATEGORY 2A SEALS Pumpable cementitious materials with no steel reinforcement and no hitching and with and without carbon fiber retrofit. Compressive strength of 200 psi and greater than 48 in thick, 4 structures tested							
Category 2A: Structure 1 Pumpable cementitious material with Blastseal Seal retrofit by Transco Mine Services Co. Weiss and Harteis [2008] Cat2A_Struct#1_2Test.xls	C drift at 320 ft from closed end 18.7 ft wide by 7.3 ft high by 5.1 ft thick Explosion tests Reflected Uniform	A 4-ft-thick pumpable cementitious seal similar to Figure 23. Celuseal material was used with a 100- to 190-psi (143-psi average) compressive strength at 36 days. A 13-in-thick retrofit structure using a solid-concrete-block wall, polyurethane, carbon fiber-reinforced polymer, and heavy steel angle iron reinforcement across the roof and floor was installed against the pumpable seal on the nonexplosion side.	508 509	No damage Destroyed	66 195.5	0.60 >5.77	A-18 A-19
Category 2A: Structure 2 Pumpable cementitious material seal R. G. Johnson Co., Inc. Sapko et al. [2005] Cat2A_Struct#2_1Test.xls	Small hydrostatic test chamber 21.2 ft wide by 8.7 ft high by 4 ft thick Methane ignition test Quasi-static Uniform	4-ft-thick pumpable cementitious foam seal similar to Figure 23. 350-psi ±75-psi compressive strength. Methane ignition test.	C3-44E	Destroyed	32	>3	A-20

Table A-1.—Detailed summary of seal structure tests—Continued

Seal category – Structure No. Description Manufacturer NIOSH reference Excel data file name	Seal test location, dimensions, test type, and loading conditions	Seal description from LLEM test files	LLEM test No.	Test outcome	Maximum pressure data (psi)	Maximum displacement data (in)	Appendix figure No.
Category 2A: Structure 3 Pumpable cementitious material >36 in R. G. Johnson Co., Inc. Sapko et al. [2005] Cat2A_Struct#3_3Test.xls	Small hydrostatic test chamber 21.2 ft wide by 8.7 ft high by 4 ft thick Water tests Quasi-static Uniform	4-ft-thick pumpable cementitious foam seal similar to Figure 23. 350-psi ±75-psi compressive strength. Hydrostatic test.	C7-64W C7-68W C7-70W	No damage No damage No damage	21 30.5 25	0.35 1.0 0.65	A-21 A-22 A-23
Category 2A: Structure 4 Pumpable cementitious material seal R. G. Johnson Co., Inc. Sapko et al. [2005] Cat2A_Struct#4_1Test.xls	Large hydrostatic test chamber 30.8 ft wide by 15.6 ft high by 4 ft thick Methane ignition test Quasi-static Uniform	4-ft-thick pumpable cementitious foam seal similar to Figure 23. 350-psi ±75-psi compressive strength. Methane ignition test.	L2-51E	Destroyed	27.5	>3	A-24
CATEGORY 2B SEALS Pumpable cementitious materials with no steel reinforcement and no hitching Compressive strength of 433 psi and greater than 36 in thick, 1 structure tested							
Category 2B: Structure 1 Pumpable cementitious material >36 in HelTech Corp. Ribfill Weiss et al. [2002] Cat2B_Struct#1_2Test.xls	B-C drifts, X-3 19.4 ft wide by 6.9 ft high by 36 in thick Explosion tests Quasi-static Nonuniform	36-in-thick pumpable cementitious seal similar to Figure 23. Approximate compressive strength of 450–500 psi.	354 355	No damage No damage	37.9 32.5	NA NA	A-25

66

Table A-1.—Detailed summary of seal structure tests—Continued

Seal category – Structure No. Description Manufacturer NIOSH reference Excel data file name	Seal test location, dimensions, test type, and loading conditions	Seal description from LLEM test files	LLEM test No.	Test outcome	Maximum pressure data (psi)	Maximum displacement data (in)	Appendix figure No.
		CATEGORY 2C SEALS Pumpable cementitious materials with no steel reinforcement and no hitching Compressive strength of 480–600 psi and 24–30 in thick, 4 structures tested					
Category 2C: Structure 1 Pumpable cementitious material 24–30 in HeiTech Corp. Hydrocrete seal Weiss et al. [2002] Cat2C_Struct#1_2Test.xls	B-C drifts, X-2 19.4 ft wide by 6.9 ft high by 34 in thick Explosion tests Quasi-static Nonuniform	34-in-thick Hydrocrete/aggregate seal similar to Figure 23. 7/8-in-thick reinforcement rods were equally spaced along ribs (2 each), floor (3), and roof (3). These bolts were grouted about 3 ft into strata with 3 ft extending into crosscut. Wood posts, cross boards, brattice, and wire screen used as forms. The aggregate was dry, crushed limestone aggregate ranging in size from 0.25 to 1 in. Compressive strength test results ranged from 600 to 840 psi.	354 355	No damage No damage	36.0 38.2	NA NA	A-26
Category 2C: Structure 2 Pumpable cementitious material 24–30 in HeiTech Corp. Hydroseal Weiss et al. [2002] Cat2C_Struct#2_2Test.xls	B-C drifts, X-4 19 ft wide by 7.55 ft high by 30 in thick Explosion tests Quasi-static Nonuniform	30-in-thick pumpable seal similar to Figure 23. Formwork consisted of wood posts, cross boards, brattice, and wire screen. Injected slurry (Hydroseal) between form walls. Compressive strength tests results ranged from 390 to 565 psi.	354 355	No damage No damage	30.1 28.5	NA NA	A-27
Category 2C: Structure 3 Pumpable cementitious material 24–30 in HeiTech Corp. Hydroseal Weiss et al. [2002] Cat2C_Struct#3_2Test.xls	B-C drifts, X-5 19.7 ft wide by 7.22 ft high by 24 in thick Explosion tests Quasi-static Nonuniform	24-in-thick pumpable seal similar to Figure 23. Formwork consisted of wood posts, cross boards, brattice, and wire screen. Injected slurry (Hydroseal) between form walls. Compressive strength test results ranged from 600 to 750 psi.	354 355	No damage No damage	30.1 28.5	NA NA	A-28

Table A-1.—Detailed summary of seal structure tests—Continued

Seal category – Structure No. Description Manufacturer NIOSH reference Excel data file name	Seal test location, dimensions, test type, and loading conditions	Seal description from LLEM test files	LLEM test No.	Test outcome	Maximum pressure data (psi)	Maximum displacement data (in)	Appendix figure No.
Category 2C: Structure 4 Pumpable cementitious material 24–30 in HeiTech Corp. rapid-construction column bag pumpable seal Weiss et al. [2002] Cat2C_Struct#4_4Test.xls	B-C drifts, X-2 19 ft wide by 6.75 ft high by 30 in thick Explosion tests Quasi-static Nonuniform	30-in-thick HeiTech grout column seal consisting of six 30-in-diam reinforced brattice bag columns spaced across the crosscut with 8- to 18-in gaps between bags and ribs. Bags were filled with 600- to 800-psi compressive strength pumpable cementitious slurry, which hardens within 10 min and is nearly fully cured in 24 hr. Similar bags without the reinforcement were then installed between the columns and filled with the same grout. 597-psi ±63-psi compressive strength after 13 days. Test 1 conducted 10 days after construction. No hitching. Polyurethane foam (Silent Seal) was sprayed around the seal perimeter and between the columns from the B-drift side.	403 404 405 406	No damage No damage No damage Damaged	20.8 28.7 26.7 35.4	1.14 2.14 0.11 >3.2	A-29 A-30 A-31 A-32

CATEGORY 3A SEALS

Articulated structures: Solid and hollow-core concrete blocks with or without hitching
Standard solid-concrete-block seal with hitching, 7 structures tested

Seal category – Structure No. Description Manufacturer NIOSH reference Excel data file name	Seal test location, dimensions, test type, and loading conditions	Seal description from LLEM test files	LLEM test No.	Test outcome	Maximum pressure data (psi)	Maximum displacement data (in)	Appendix figure No.
Category 3A: Structure 1 Standard seal with hitching NIOSH constructed Sapko et al. [2005] Cat3A_Struct#1_4Test.xls	B-C drifts, X-1 18.3 ft wide by 6.7 ft high by 16 in thick Explosion tests Quasi-static Nonuniform	Standard solid-concrete-block seal without pilaster similar to Figure 26. Simulated rib and floor hitching with 6-in by 6-in by 0.5-in-thick steel angle anchored on 18-in centers using 1-in-diam by 9-in-long Hilti Kwik Bolt III. Fully mortared and staggered joints.	403 404 405 406	No damage No damage No damage No damage	18.1 22.4 28.3 30.8	0.00 0.00 0.00 0.10	A-33 A-34 A-35 A-36

Table A-1.—Detailed summary of seal structure tests—Continued

Seal category – Structure No. Description Manufacturer NIOSH reference Excel data file name	Seal test location, dimensions, test type, and loading conditions	Seal description from LLEM test files	LLEM test No.	Test outcome	Maximum pressure data (psi)	Maximum displacement data (in)	Appendix figure No.
Category 3A: Structure 2 Standard seal with hitching NIOSH constructed Not previously reported by NIOSH Cat3A_Struct#2_10Test.xls	B-C drifts, X-1 18.3 ft wide by 6.7 ft high by 16 in thick Explosion tests Quasi-static Nonuniform	Standard solid-concrete-block seal with 16-in-wide by 32-in-thick center pilaster identical to Figure 26. Simulated rib and floor hitching with 6-in by 6-in by 0.5-in-thick steel angle anchored on 18-in centers using 1-in-diam by 9-in-long Hilti Kwik Bolt III. Quikrete BlocBond (1225-51) was used as mortar.	500 501 502 503 504 505 506 507 508 509	No damage No damage No damage No damage No damage No damage No damage No damage No damage No damage	23 22 22 15 16 24 34 24 22 46	NA NA NA NA NA NA NA NA NA NA	A-37 A-38
Category 3A: Structure 3 Standard seal with hitching NIOSH constructed Not previously reported by NIOSH Cat3A_Struct#3_2Test.xls	B-C drifts, X-3 18.6 ft wide by 6.8 ft high by 16 in thick Explosion tests Quasi-static Nonuniform	Standard solid-concrete-block seal with 16-in-wide by 32-in-thick center pilaster identical to Figure 26. Used Type S mortar and B-bond face coatings. Simulated rib and floor hitching with 6-in by 6-in by 0.5-in-thick steel angle anchored on 18-in centers using 1-in-diam by 9-in-long Hilti Kwik Bolt III.	506 507	No damage No damage	63 30	0.11 0.07	A-39 A-40
Category 3A: Structure 4 Standard seal with hitching NIOSH constructed Sapko et al. [2005] Cat3A_Struct#4_5Test.xls	Small hydrostatic test chamber 16.8 ft wide by 8.6 ft high by 16 in thick Methane ignition tests Quasi-static Uniform	Standard solid-concrete-block seal identical to Figure 26. 2,500 psi (±100-psi) concrete blocks. 16-in-wide by 32-in-thick pilaster. Simulated rib and floor hitching with 6-in by 6-in by 0.5-in-thick steel angle anchored on 18-in centers using 1-in-diam by 9-in-long Hilti Kwik Bolt III. Methane ignition tests.	C1-5E C1-8E C1-9E C1-10E C1-11E	No damage No damage No damage No damage Destroyed	56.5 90.1 94.3 79.5 99.7	0.13 NA 0.22 0.21 >2.8	A-41 A-42 A-43 A-44 A-45

Table A-1.—Detailed summary of seal structure tests—Continued

Seal category – Structure No. Description Manufacturer NIOSH reference Excel data file name	Seal test location, dimensions, test type, and loading conditions	Seal description from LLEM test files	LLEM test No.	Test outcome	Maximum pressure data (psi)	Maximum displacement data (in)	Appendix figure No.
Category 3A: Structure 5 Standard seal with hitching NIOSH constructed Sapko et al. [2005] Cat3A_Struct#5_2Test.xls	Small hydrostatic test chamber 16.8 ft wide by 8.6 ft high by 16 in thick Water and methane ignition tests Quasi-static Uniform	Standard solid-concrete-block seal identical to Figure 26. 2,500-psi (±100-psi) concrete blocks. 16-in-wide by 32-in-thick pilaster. Simulated rib and floor hitching with 6-in by 6-in by 0.5-in-thick steel angle anchored on 18-in centers using 1-in-diam by 9-in-long Hilti Kwik Bolt III. Methane ignition and hydrostatic tests.	C6-60W C6-62E	No damage No damage	32 88	0.024 0.25	A-46 A-47
Category 3A: Structure 6 Standard seal with hitching NIOSH constructed Sapko et al. [2005] Cat3A_Struct#6_1Test.xls	Large hydrostatic test chamber 28 ft wide by 15.5 ft high by 16 in thick Methane ignition test Quasi-static Uniform	Standard solid-concrete-block seal identical to Figure 26. 2,500-psi (±100-psi) concrete blocks. 16-in-wide by 32-in-thick pilaster. Simulated rib and floor hitching with 6-in by 6-in by 0.5-in-thick steel angle anchored on 18-in centers using 1-in-diam by 9-in-long Hilti Kwik Bolt III. Methane ignition test.	L1-37E	Destroyed	32	>3	A-48
Category 3A: Structure 7 Standard seal with hitching NIOSH constructed Sapko et al. [2005] Cat3A_Struct#7_1Test.xls	SRCM hydrostatic test chamber 18 ft wide by 6.2 ft high by 16 in thick Water test Quasi-static Uniform	Standard solid-concrete-block seal identical to Figure 26. 1,900- to 2,500-psi (±100-psi) concrete blocks. 16-in-wide by 32-in-thick pilaster. Simulated rib and floor hitching with 6-in by 6-in by 0.5-in-thick steel angle anchored on 18-in centers using 1-in-diam by 9-in-long Hilti Kwik Bolt III. Hydrostatic test.	SRCM-1	Destroyed	20	0.15	A-49

Table A-1.—Detailed summary of seal structure tests—Continued

Seal category – Structure No. Description Manufacturer NIOSH reference Excel data file name	Seal test location, dimensions, test type, and loading conditions	Seal description from LLEM test files	LLEM test No.	Test outcome	Maximum pressure data (psi)	Maximum displacement data (in)	Appendix figure No.
CATEGORY 3B SEALS							
Articulated structures: Solid and hollow-core concrete blocks with or without hitching Solid-concrete-block seal with Packsetter Bags and without hitching, 3 structures tested							
Category 3B: Structure 1 Standard seal with grout bags Strata Mine Services Packsetter seals Weiss et al. [2002] Cat3B_Struct#1_2Test.xls	B-C drifts, X-2 19 ft wide by 6.9 ft high by 16 in thick Explosion tests Quasi-static Nonuniform	Standard solid-concrete-block seal similar to Figure 30 with center pilaster and floor hitching simulated by steel angle. Used tongue-and-groove, 8-in solid blocks with fully mortared joints and face coatings. Packsetter Bags from Strata Products used on roof and ribs to eliminate need for rib hitching in friable rib conditions. Packsetter Bags pressured to 44 psi. Compressive strength of grout in bags is 360 psi after 1 day, 435 psi after 7 days, and 580 psi after 28 days.	365 366	No damage No damage	33.9 —	NA NA	A-50
Category 3B: Structure 2 Standard seal with grout bags Strata Mine Services Packsetter seals Weiss et al. [2002] Cat3B_Struct#2_2Test.xls	B-C drifts, X-3 19.4 ft wide by 6.9 ft high by 16 in thick Explosion tests Quasi-static Nonuniform	Standard solid-concrete-block seal identical to Figure 30 with center pilaster. No floor hitching simulated by steel angle. Used tongue-and-groove, 8-in solid blocks with fully mortared joints and face coatings. Packsetter Bags from Strata Products used on roof and ribs to eliminate need for rib hitching in friable rib conditions. Packsetter Bags pressured to 36–40 psi. Compressive strength of grout in bags is 360 psi after 1 day, 435 psi after 7 days, and 580 psi after 28 days.	365 366	No damage No damage	49.5 40.4	NA NA	A-51
Category 3B: Structure 3 Standard seal with grout bags Strata Mine Services Packsetter seals Weiss et al. [2002] Cat3B_Struct#3_2Test.xls	B-C drifts, X-4 19 ft wide by 7.2 ft high by 16 in thick Explosion tests Quasi-static Nonuniform	Standard solid-concrete-block seal with center pilaster similar to Figure 30. No floor hitching simulated by steel angle. Used tongue-and-groove, 8-in solid blocks that were dry-stacked with face coatings. Packsetter Bags from Strata Products used on roof and ribs to eliminate need for rib hitching in friable rib conditions. Packsetter Bags pressured to 44 psi. Compressive strength of grout in bags is 360 psi after 1 day, 435 psi after 7 days, and 580 psi after 28 days.	365 366	No damage Destroyed	31.7 41.6	NA NA	A-52

Table A-1.—Detailed summary of seal structure tests—Continued

Seal category – Structure No. Description Manufacturer NIOSH reference Excel data file name	Seal test location, dimensions, test type, and loading conditions	Seal description from LLEM test files	LLEM test No.	Test outcome	Maximum pressure data (psi)	Maximum displacement data (in)	Appendix figure No.
CATEGORY 3C SEALS							
Articulated structures: Solid and hollow-core concrete blocks with or without hitching Ventilation stoppings: Solid and hollow-core concrete blocks without hitching, 8 structures tested							
Category 3C: Structure 1 Ventilation stopping: Hollow-core concrete block NIOSH constructed Weiss et al. [2008] Cat3C_Struct#1_2Test.xls	B-C drifts, X-4 18.9 ft wide by 7.1 ft high by 6 in thick Explosion tests Quasi-static Nonuniform	Dry-stacked, staggered-joint, hollow-core concrete block stopping identical to Figure 32 on small concrete foundation located about 5 ft from the center of the crosscut toward B-drift. Mortar used to level the first block course. 3-core, concrete block with nominal dimensions of 6 in by 8 in by 16 in and 1,000-psi compressive strength. Wood wedges used to tighten each course at the ribs and wood header boards and wedges were used across the interface of the top course to the mine roof to tighten the stopping. Quikrete BlocBond sealant (No. 1227-50; MSHA IC 36) applied to both faces of each stopping.	427 428	No damage Destroyed	0.60 4.0	0.00 >3.5	A-53 A-54
Category 3C: Structure 2 Ventilation stopping: Hollow-core concrete block NIOSH constructed Weiss et al. [2008] Cat3C_Struct#2_2Test.xls	B-C drifts, X-5 18.8 ft wide by 7.3 ft high by 6 in thick Explosion tests Quasi-static Nonuniform	Dry-stacked, staggered-joint, hollow-core concrete block stopping identical to Figure 32 on small concrete foundation located about 5 ft from the center of the crosscut toward B-drift. Mortar used to level the first block course. 3-core, concrete block with nominal dimensions of 6 in by 8 in by 16 in and 1,000-psi compressive strength. Wood wedges used to tighten each course at the ribs and wood header boards and wedges were used across the interface of the top course to the mine roof to tighten the stopping. Quikrete BlocBond sealant (No. 1227-50; MSHA IC 36) applied to both faces of each stopping.	427 428	No damage Destroyed	0.60 4.0	0.00 >3.5	A-55 A-56

Table A-1.—Detailed summary of seal structure tests—Continued

Seal category – Structure No. Description Manufacturer NIOSH reference Excel data file name	Seal test location, dimensions, test type, and loading conditions	Seal description from LLEM test files	LLEM test No.	Test outcome	Maximum pressure data (psi)	Maximum displacement data (in)	Appendix figure No.
Category 3C: Structure 3 Ventilation stopping: Hollow-core concrete block NIOSH constructed Weiss et al. [2008] Cat3C_Struct#3_6Test.xls	B-C drifts, X-6 17.2 ft wide by 7.1 ft high by 6 in thick Explosion tests Quasi-static Nonuniform	Dry-stacked, staggered-joint, hollow-core concrete block stopping identical to Figure 32 on small concrete foundation located about 5 ft from the center of the crosscut toward B-drift. Mortar used to level the first block course. 3-core, concrete block with nominal dimensions of 6 in by 8 in by 16 in with 1,000-psi compressive strength. Wood wedges used to tighten each course at the ribs and wood header boards and wedges were used across the interface of the top course to the mine roof to tighten the stopping. Quikrete BlocBond sealant (No. 1227-50; MSHA IC 36) applied to both faces of each stopping.	427 428 429 430 432 433	No damage No damage No damage Damaged Damaged Destroyed	0.50 3.2 1.2 3.2 1.8 3.2	0.00 0.01 0.15 1.0 1.5 >4.0	A-57 A-58 A-59 A-60 A-61 A-62
Category 3C: Structure 4 Ventilation stopping: Hollow-core concrete block NIOSH constructed Weiss et al. [2008] Cat3C_Struct#4_7Test.xls	B-C drifts, X-7 16.8 ft wide by 7.3 ft high by 6 in thick Explosion tests Quasi-static Nonuniform	Dry-stacked, staggered-joint, hollow-core concrete block stopping identical to Figure 32 on small concrete foundation located about 5 ft from the center of the crosscut toward B-drift. Mortar used to level the first block course. 3-core, concrete block with nominal dimensions of 6 in by 8 in by 16 in with 1,000-psi compressive strength. Wood wedges used to tighten each course at the ribs and wood header boards and wedges were used across the interface of the top course to the mine roof to tighten the stopping. Quikrete BlocBond sealant (No. 1227-50; MSHA IC 36) applied to both faces of each stopping.	427 428 429 430 432 433 434	No damage No damage No damage No damage No damage No damage Damaged	0.6 3.5 1.0 2.6 1.5 3.0 2.5	0.04 0.19 0.13 0.34 0.43 0.68 0.93	A-63 A-64 A-65 A-66 A-67 A-68 A-69
Category 3C: Structure 5 Ventilation stopping: Solid concrete block NIOSH constructed Weiss et al. [2008] Cat3C_Struct#5_6Test.xls	B-C drifts, X-4 18.9 ft wide by 7.1 ft high by 6 in thick Explosion tests Quasi-static Nonuniform	Dry-stacked, staggered-joint, solid-concrete-block stopping identical to Figure 32 on small concrete foundation located about 5 ft from the center of the crosscut toward B-drift. Mortar used to level the first block course. Standard solid concrete block with nominal dimensions of 6 in by 8 in by 16 in and 1,300- to 1,740-psi compressive strength. Wood wedges used to tighten each course at the ribs and wood header boards and wedges were used across the interface of the top course to the mine roof to tighten the stopping. Quikrete BlocBond sealant (No. 1227-50; MSHA IC 36) applied to both faces of each stopping.	457 458 459 460 461 462	No damage No damage No damage Damaged Damaged Destroyed	0.80 1.60 4.7 7.5 4.2 7.7	0.00 0.06 0.67 1.15 0.60 >3.00	A-70 A-71 A-72 A-73 A-74 A-75

Table A-1.—Detailed summary of seal structure tests—Continued

Seal category – Structure No. Description Manufacturer NIOSH reference Excel data file name	Seal test location, dimensions, test type, and loading conditions	Seal description from LLEM test files	LLEM test No.	Test outcome	Maximum pressure data (psi)	Maximum displacement data (in)	Appendix figure No.
Category 3C: Structure 6 Ventilation stopping: Solid concrete block NIOSH constructed Weiss et al. [2008] Cat3C_Struct#6_7Test.xls	B-C drifts, X-5 18.8 ft wide by 7.3 ft high by 6 in thick Explosion tests Quasi-static Nonuniform	Dry-stacked, staggered-joint, solid-concrete-block stopping identical to Figure 32 on small concrete foundation located about 5 ft from the center of the crosscut toward B-drift. Mortar used to level the first block course. Standard solid concrete block with nominal dimensions of 6 in by 8 in by 16 in and 1,300- to 1,740-psi compressive strength). Wood wedges used to tighten each course at the ribs and wood header boards and wedges were used across the interface of the top course to the mine roof to tighten the stopping. Quikrete BlocBond sealant (No. 1227-50; MSHA IC 36) applied to both faces of each stopping.	457 458 459 460 461 462 463	No damage No damage No damage No damage No damage No damage Destroyed	0.75 1.4 3.7 5.4 3.9 6.7 19.5	0.00 0.05 0.41 0.76 0.32 0.93 >3.2	A-76 A-77 A-78 A-79 A-80 A-81 A-82
Category 3C: Structure 7 Ventilation stopping: Solid concrete block NIOSH constructed Weiss et al. [2008] Cat3C_Struct#7_10Test.xls	A-B drifts, X-6 19.1 ft wide by 7.6 ft high by 8 in thick Explosion tests Quasi-static Nonuniform	8-in-thick, solid-concrete-block stopping identical to Figure 32 with fully mortared (Type S) joints. Quikrete B-Bond (No. 1227-50) applied 0.25 in thick on A-drift face. This stopping is located approximately 10 ft into the crosscut as measured from the A-drift rib line.	510 511 512 513 514 515 516 517 518 519	No damage No damage No damage No damage No damage No damage No damage No damage No damage No damage	16.8 12.1 12.2 11.3 18.1 14.8 19.6 16.1 13.7 41.1	NA NA NA NA NA NA NA NA NA NA	A-83 A-84 A-85
Category 3C: Structure 8 Ventilation stopping: Solid concrete block NIOSH constructed Weiss et al. [2008] Cat3C_Struct#8_10Test.xls	A-B drifts, X-7 19.5 ft wide by 7.4 ft high by 6 in thick Explosion tests Quasi-static Nonuniform	6-in-thick, solid-concrete-block stopping identical to Figure 32 with fully mortared (Type S) joints. Quikrete B-Bond (No. 1227-50) applied 0.25 in thick on A-drift face. This stopping is located approximately 10 ft into the crosscut as measured from the A-drift rib line.	510 511 512 513 514 515 516 517 518 519	No damage No damage No damage No damage No damage No damage No damage No damage No damage Destroyed	19.1 11.0 11.2 11.3 12.2 17.5 14.4 15.3 12.0 28.8	0.51 0.41 0.45 0.47 0.54 0.72 0.68 0.74 0.74 2.76	A-86 A-87 A-88 A-89 A-90 A-91 A-92 A-93 A-94 A-95

Table A-1.—Detailed summary of seal structure tests—Continued

Seal category – Structure No. Description Manufacturer NIOSH reference Excel data file name	Seal test location, dimensions, test type, and loading conditions	Seal description from LLEM test files	LLEM test No.	Test outcome	Maximum pressure data (psi)	Maximum displacement data (in)	Appendix figure No.
CATEGORY 4 SEAL Polymer and aggregate materials without hitching, 1 structure tested							
Category 4: Structure 1 Polyurethane foam and aggregate seal NIOSH constructed Not previously reported by NIOSH Cat4_Struct#1_1Test.xls	Small hydrostatic test chamber 20.4 ft wide by 9 ft high by 11 in thick Methane ignition test Quasi-static Uniform	Polyurethane foam and aggregate seal similar to Figure 35.	C8	Destroyed	19	0.5	A-96, A-97, A-98, A-99
CATEGORY 5A SEALS Wood-crib-block seals with or without hitching Wood-crib-block seal with hitching, 0 structures tested							
Category 5A Wood-crib-block seal with hitching	—	No tests reported here. See Figure 38 for typical plan.	—	—	—	—	—

75

Table A-1.—Detailed summary of seal structure tests—Continued

Seal category – Structure No. Description Manufacturer NIOSH reference Excel data file name	Seal test location, dimensions, test type, and loading conditions	Seal description from LLEM test files	LLEM test No.	Test outcome	Maximum pressure data (psi)	Maximum displacement data (in)	Appendix figure No.
CATEGORY 5B SEALS							
		Wood-crib-block seal with or without hitching Wood-crib-block seal with glue and Packsetter Bags, 1 structure tested					
Category 5B: Structure 1 Wood-crib-block seal with glue and Packsetter Bags Strata Mine Services wood block seal with Packsetter Bags Sapko et al. [2003] Cat5B_Struct#1_2Test.xls	B-C drifts, X-1 18.4 ft wide by 6.5 ft high by 30 in thick Explosion tests Quasi-static Nonuniform	30-in-thick seal identical to Figure 41 using 5-in by 6-in by 30-in-long Eastern Oak crib blocks. The wood blocks were installed on a level concrete pad lengthwise parallel with the crosscut ribs. Alternating courses with the 5- or 6-in dimensions of the crib in the vertical dimension to achieve a staggered joint pattern. Approximately 36–40 blocks per course, 14 courses high and about 550 blocks total. 3 beads of adhesive applied between the blocks and block courses. 12 Packsetter Bags installed and injected with grout to about 60 psi along the mine roof and ribs. Compressive strength of grout in bags is 360 psi after 1 day, 435 psi after 7 days, and 580 psi after 28 days. Hitching along the floor and ribs was not used. Polyurethane foam was injected into the gaps between the Packsetter Bags and the mine strata/block interface. Each side was spray-coated with Eagle sealant; brattice curtain was installed across the B-drift face and sprayed with the sealant. Tested 14 days after construction.	396 399	No damage No damage	23.2 32.4	0.79 1.9	A-100 A-101
CATEGORY 6A SEALS							
		Articulated structures: Lightweight blocks with or without hitching Lightweight blocks: 24 in thick with hitching, 2 structures tested					
Category 6A: Structure 1 Omega blocks with hitching Burrell Mining Products Omega block seal Cashdollar et al. [2007] Cat6A_Struct#1_2Test.xls	B-C drifts, X-3 18.8 ft wide by 6.75 ft high by 24 in thick Explosion tests Quasi-static Nonuniform	24-in-thick Omega block seal identical to Figure 44 with fully mortared joints (Quikrete BlocBond), a 48-in by 48-in interlocked center pilaster, and simulated hitch into the ribs and floor with 6-in by 6-in by 0.5-in-thick steel angle bolted to the ribs and floor.	508 509	No damage Damaged	52.9 70	0.04 2.2	A-102 A-103

Table A-1.—Detailed summary of seal structure tests—Continued

Seal category – Structure No. Description Manufacturer NIOSH reference Excel data file name	Seal test location, dimensions, test type, and loading conditions	Seal description from LLEM test files	LLEM test No.	Test outcome	Maximum pressure data (psi)	Maximum displacement data (in)	Appendix figure No.
Category 6A: Structure 2 Omega blocks with hitching Burrell Mining Products Omega block seal Sapko et al. [2005] Cat6A_Struct#2_1Test.xls	Small hydrostatic test chamber 20 ft wide by 8.2 ft high by 24 in thick Methane ignition test Quasi-static Uniform	Properly constructed, Omega block seal identical to Figure 44 with fully mortared joints (Quikrete BlocBond), a 48-in by 48-in interlocked center pilaster, and simulated hitch into the ribs and floor with 6-in by 6-in by 0.5-in-thick steel angle bolted to the ribs and floor. 100-psi ±20-psi compressive strength. Methane ignition test.	C4-48E	Destroyed	22	>3	A-104

CATEGORY 6B SEALS

Articulated structures: Lightweight blocks with or without hitching
Lightweight blocks: 40 in thick without hitching, 8 structures tested

Seal category – Structure No. Description Manufacturer NIOSH reference Excel data file name	Seal test location, dimensions, test type, and loading conditions	Seal description from LLEM test files	LLEM test No.	Test outcome	Maximum pressure data (psi)	Maximum displacement data (in)	Appendix figure No.
Category 6B: Structure 1 Omega blocks, no hitching Burrell Mining Products rapid-design Omega block seal Sapko et al. [2003] Cat6B_Struct#1_4Test.xls	B-C drifts, X-3 19 ft wide by 6.75 ft high by 40 in thick Explosion tests Quasi-static Nonuniform	Burrell's 40-in-thick Omega 384 block seal identical to Figure 48 with fully mortared and staggered joints and full face coatings using Quikrete BlocBond. Rough-cut lumber was used across the top of the seal and then wedged tightly at the mine roof. No hitching.	403	No damage	23.4	0.29	A-105
			404	No damage	29.7	0.49	A-106
			405	No damage	30.2	0.06	A-107
			406	No damage	31.2	0.78	A-108
Category 6B: Structure 2 Omega blocks, no hitching Burrell Mining Products Omega block seal Cashdollar et al. [2007] Cat6B_Struct#2_9Test.xls	B-C drifts, X-2 18.8 ft wide by 6.7 ft high by 40 in thick Explosion tests Quasi-static Nonuniform	Properly built, 40-in-thick Omega block seal identical to Figure 48 as built by Burrell with fully mortared and staggered joints and full face coatings using Quikrete BlocBond. Fully mortared at top of seal. No hitching.	501	No damage	24.8	0.04	A-109
			502	No damage	23.4	0.03	A-110
			503	No damage	13.3	0.01	A-111
			504	No damage	15.1	0.01	A-112
			505	No damage	38.8	0.03	A-113
			506	No damage	67.6	0.08	A-114
			507	No damage	29.7	0.055	A-115
			508	No damage	29.1	NA	A-116
			509	No damage	118.4	NA	

Table A-1.—Detailed summary of seal structure tests—Continued

Seal category – Structure No. Description Manufacturer NIOSH reference Excel data file name	Seal test location, dimensions, test type, and loading conditions	Seal description from LLEM test files	LLEM test No.	Test outcome	Maximum pressure data (psi)	Maximum displacement data (in)	Appendix figure No.
Category 6B: Structure 3 Omega blocks, no hitching Hybrid Omega block seal Cashdollar et al. [2007] Cat6B_Struct#3_2Test.xls	B-C drifts, X-3 18.8 ft wide by 6.75 ft high by 40 in thick Explosion tests Quasi-static Nonuniform	Hybrid 40-in-thick Omega block seal similar to Figure 48. Dry layer of BlocBond applied to floor, then dampened before installing first block course. BlocBond was applied to top of first course and forced into vertical joints by hand; other courses installed in similar manner. Wedged tight (skin to skin) across top center boards (~18-in gap between centered board ends and each rib); attempts by hand to fill all gaps with BlocBond.	501 502	No damage Destroyed	28.5 51.1	0.13 >5.8	A-117 A-118
Category 6B: Structure 4 Omega blocks, no hitching Omega block seal as built at Sago Mine Cashdollar et al. [2007] Cat6B_Struct#4_3Test.xls	B-C drifts, X-3 18.8 ft wide by 6.75 ft high by 40 in thick Explosion tests Quasi-static Nonuniform	Hybrid 40-in-thick Omega block seal similar to Figure 48. Dry layer of BlocBond applied to floor, then dampened before installing first block course. BlocBond was applied to top of first course and forced into vertical joints by hand; other courses installed in similar manner. Wedged tight (skin to skin) across top rib-to-rib boards; attempts by hand to fill all gaps with BlocBond.	503 504 505	No damage No damage Destroyed	16.0 18.4 42.4	0.04 0.085 >5.4	A-119 A-120 A-121
Category 6B: Structure 5 Omega blocks, no hitching Burrell Mining Products Omega block seal Cashdollar et al. [2007] Cat6B_Struct#5_1Test.xls	C drift at 320 ft 18.7 ft wide by 7.3 ft high by 41 in thick Explosion test Reflected Uniform	Properly built, 40-in-thick Omega block seal identical to Figure 48 with staggered and fully mortared joints using BlocBond. Wedged tight across entire roof. Wedged tight (skin to skin) across top rib-to-rib boards; attempts by hand to fill all gaps with BlocBond. No hitching.	502	Destroyed	61	>6.1	A-122

Table A-1.—Detailed summary of seal structure tests—Continued

Seal category – Structure No. Description Manufacturer NIOSH reference Excel data file name	Seal test location, dimensions, test type, and loading conditions	Seal description from LLEM test files	LLEM test No.	Test outcome	Maximum pressure data (psi)	Maximum displacement data (in)	Appendix figure No.
Category 6B: Structure 6 Omega blocks, no hitching Omega block seal as built at Sago Mine Cashdollar et al. [2007] Cat6B_Struct#6_3Test.xls	C drift at 320 ft 18.7 ft wide by 7.3 ft high by 40 in thick Explosion tests Reflected Uniform	Hybrid 40-in-thick Omega block seal similar to Figure 48. Dry layer of BlocBond applied to floor, then dampened before installing first block course. BlocBond was applied to top of first course and forced into vertical joints by hand; other courses installed in similar manner. Wedged tight (skin to skin) across top rib-to-rib boards; attempts by hand to fill all gaps with BlocBond.	503 504 505	No damage No damage Destroyed	17.2 20.7 63.2	0.045 0.065 >6.0	A-123 A-124 A-125
Category 6B: Structure 7 Omega blocks, no hitching Omega block seal as built at Sago Mine using blocks from Sago Cashdollar et al. [2007] Cat6B_Struct#7_1Test.xls	C drift at 320 ft 18.7 ft wide by 7.3 ft high by 40 in thick Explosion test Reflected Uniform	Hybrid 40-in-thick Omega block seal similar to Figure 48. Dry layer of BlocBond applied to floor, then dampened before installing first block course. BlocBond was applied to top of first course and forced into vertical joints by hand; other courses installed in similar manner. Wedged tight (skin to skin) across top rib-to-rib boards; attempts by hand to fill all gaps with BlocBond.	506	Destroyed	118.2	>6.2	A-126
Category 6B: Structure 8 Omega blocks, no hitching Burrell Mining Products rapid-design Omega block Seal Sapko et al. [2005] Cat6B_Struct#8_1Test.xls	Small hydrostatic test chamber 20.6 ft wide by 8.8 ft high by 40 in thick Methane ignition test Quasi-static Uniform	Omega block seal identical to Figure 48. 100-psi ±20-psi compressive strength.	C5-53E	Destroyed	22.5	>3	A-127

NA Not available.

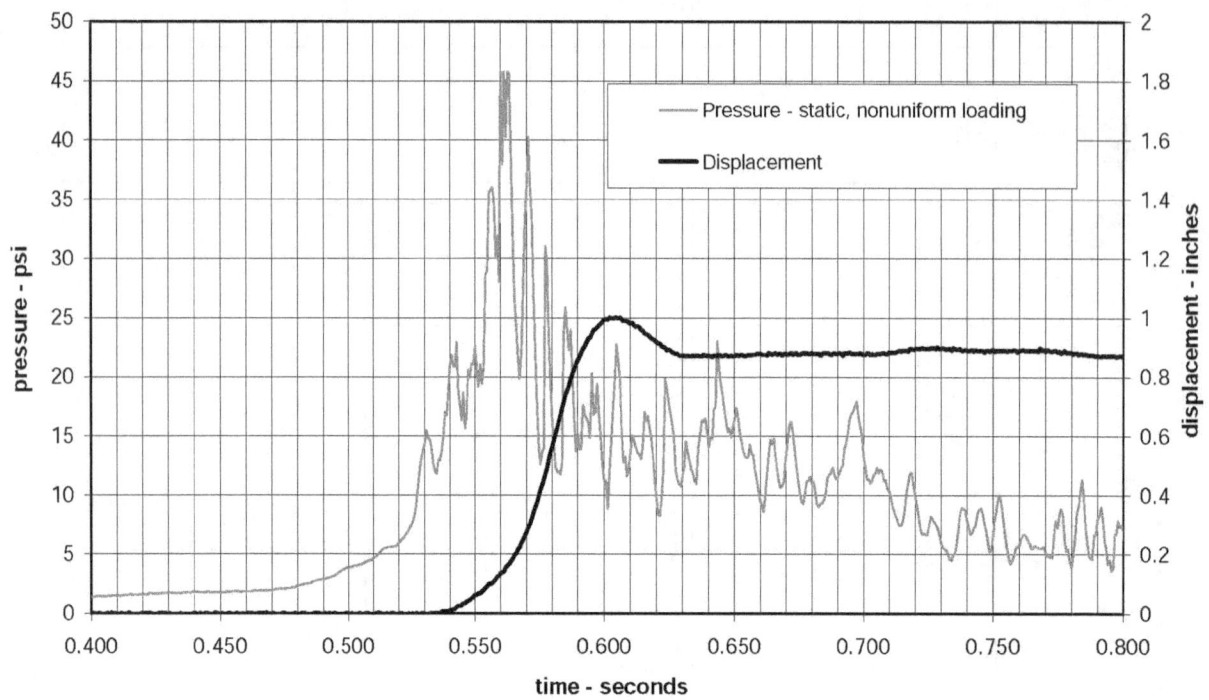

Figure A-1.—Category 1A - structure #1 - test 1 - static, nonuniform loading. Insteel 3-D seal - shotcrete with reinforcement - LLEM test #419.

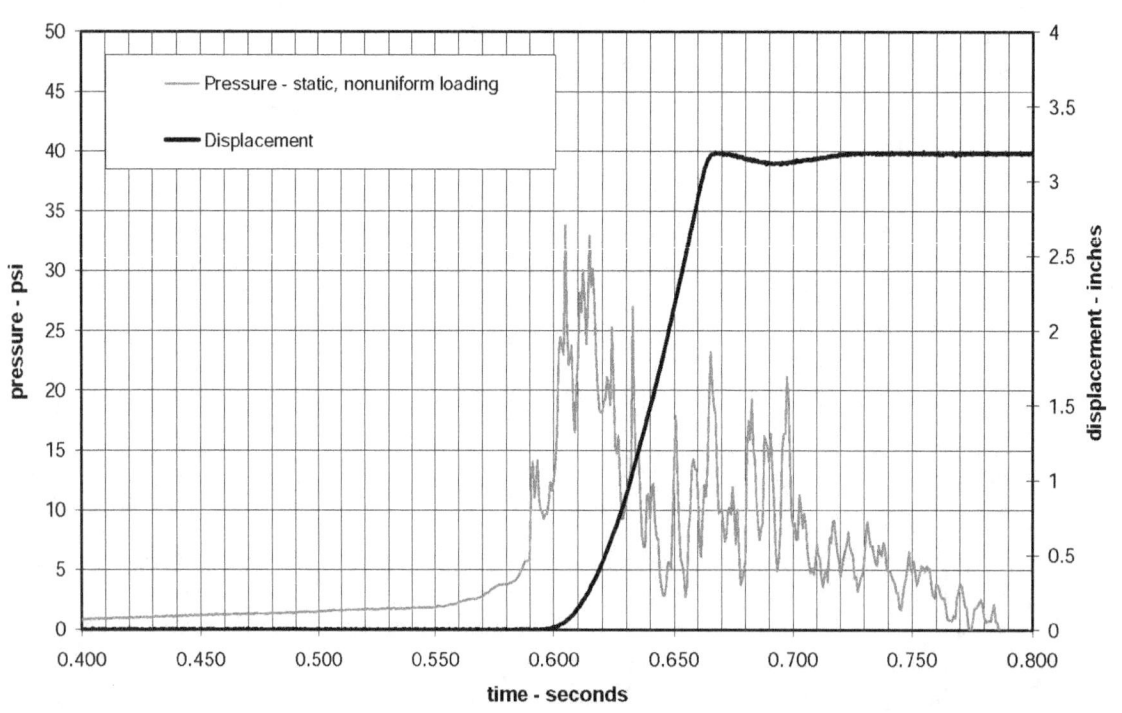

Figure A-2.—Category 1A - structure #2 - test 1 - static, nonuniform loading. Insteel 3-D seal - shotcrete with reinforcement - LLEM test #419.

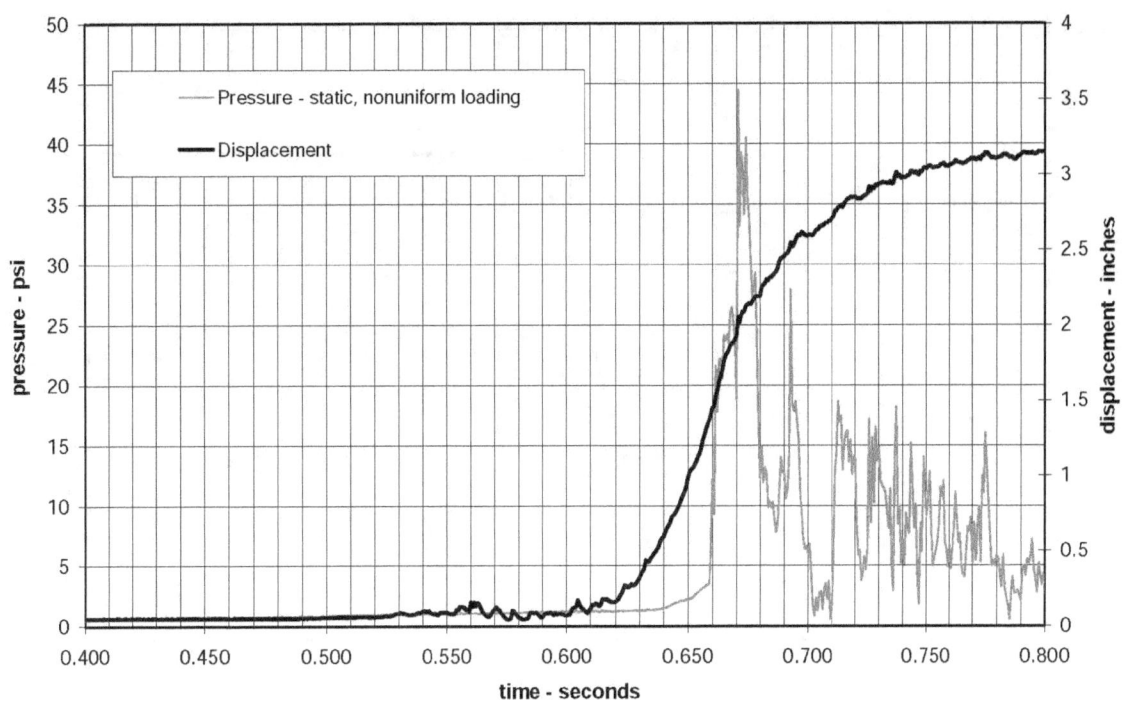

Figure A-3.—Category 1A - structure #3 - test 1 - static, nonuniform loading. Insteel 3-D seal - shotcrete with reinforcement - LLEM test #419.

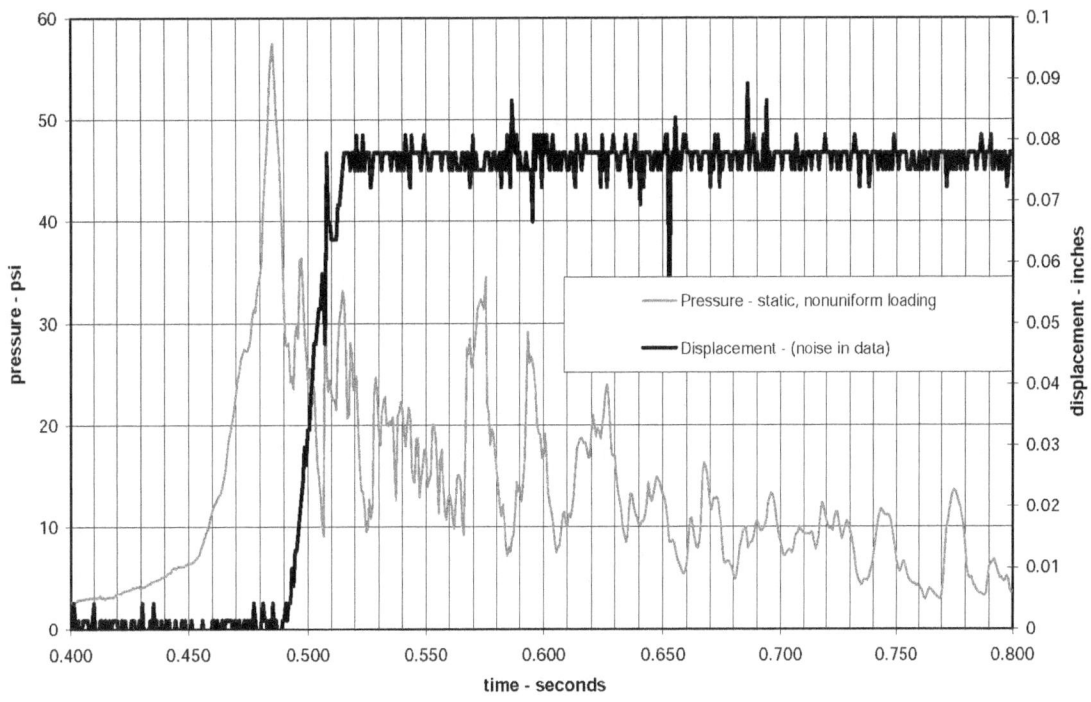

Figure A-4.—Category 1A - structure #4 - test 1 - static, nonuniform loading. Insteel 3-D seal - shotcrete with reinforcement - LLEM test #420.

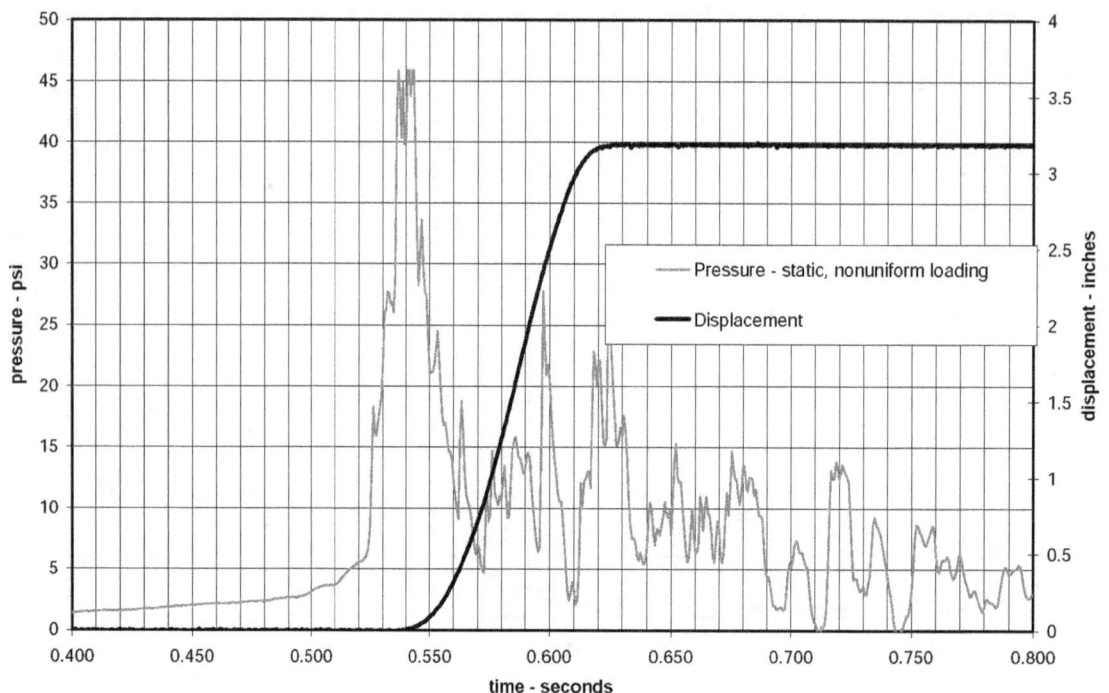

Figure A-5.—Category 1A - structure #5 - test 1 - static, nonuniform loading.
Insteel 3-D seal - shotcrete with reinforcement - LLEM test #420.

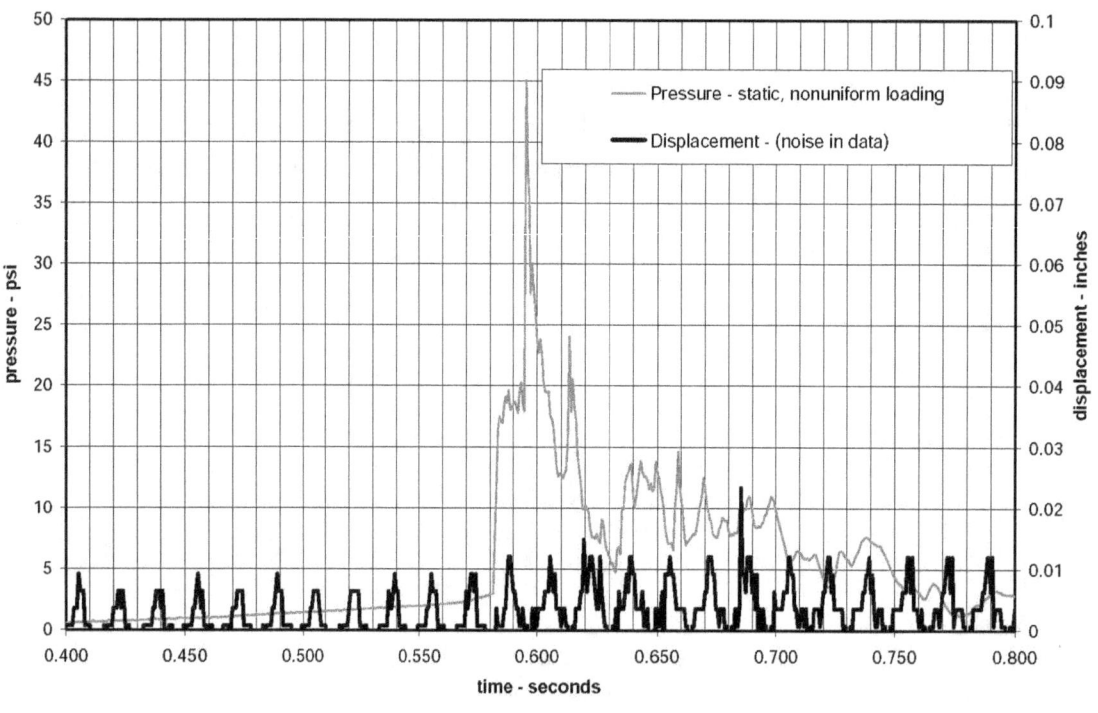

Figure A-6— Category 1A - structure #6 - test 1 - static, nonuniform loading.
Insteel 3-D seal - shotcrete with reinforcement - LLEM test #420.

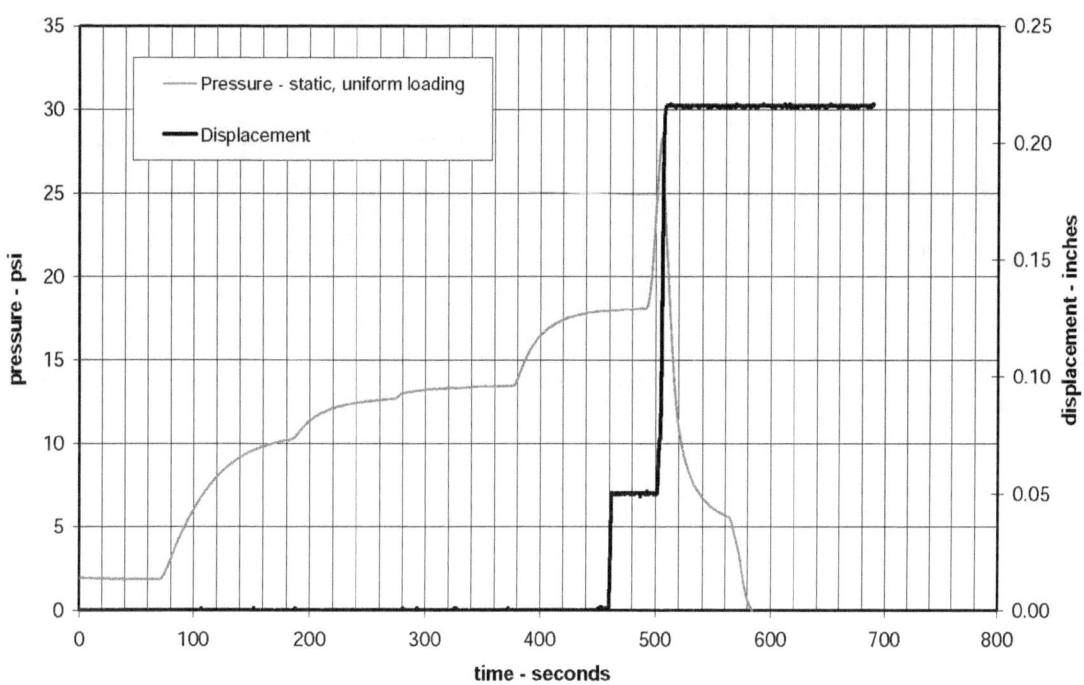

Figure A-7.—Category 1A - structure #7 - test 1 - static, uniform loading.
Insteel 3-D seal - shotcrete with reinforcement - PR-1.

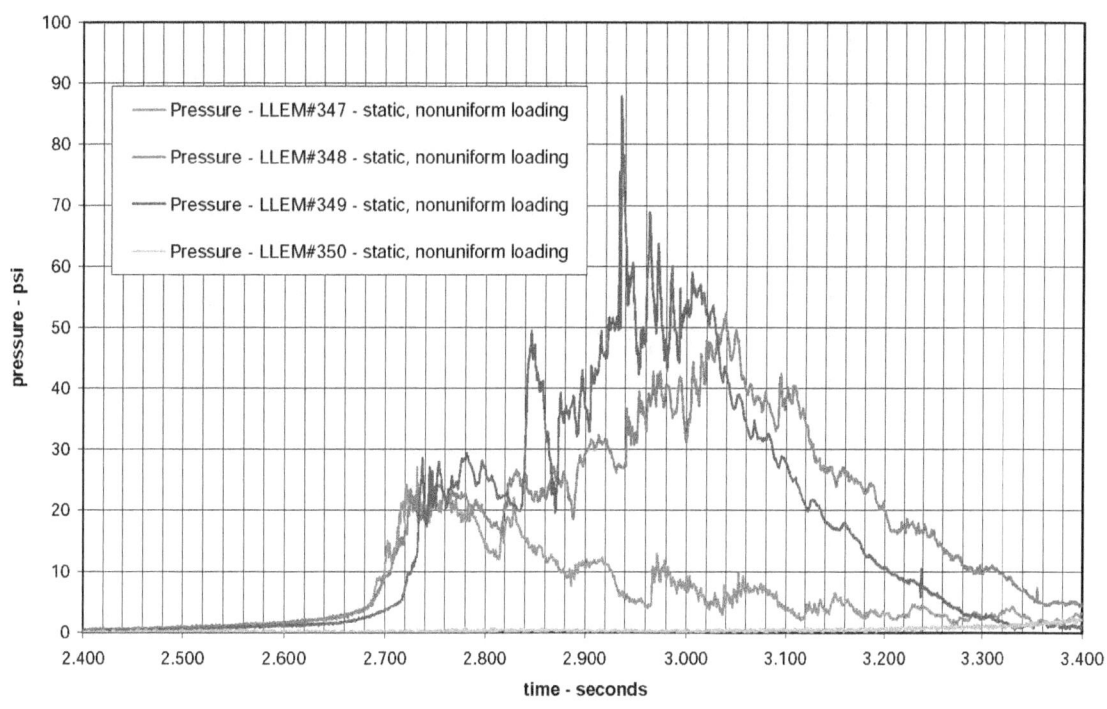

Figure A-8.—Category 1B - structure #1 - tests 1 to 4 - static, nonuniform loading.
Meshblock seal - shotcrete with reinforcement - LLEM tests #347–350.

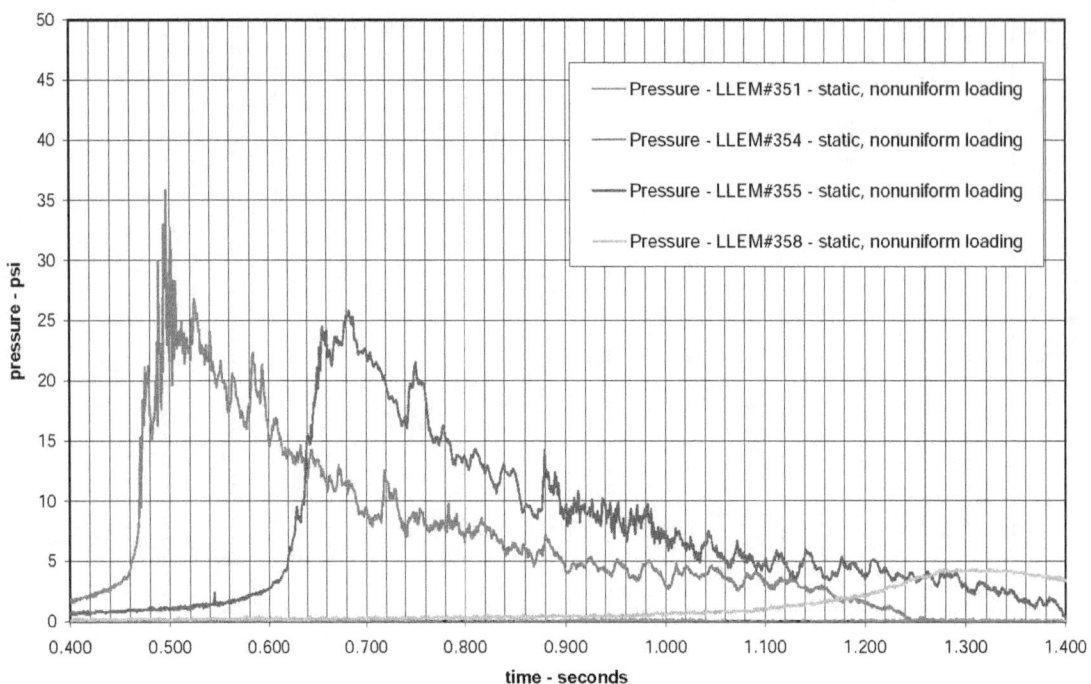

Figure A-9.—Category 1B - structure #1 - tests 5 to 8 - static, nonuniform loading. Meshblock seal - shotcrete with reinforcement - LLEM tests #351–358.

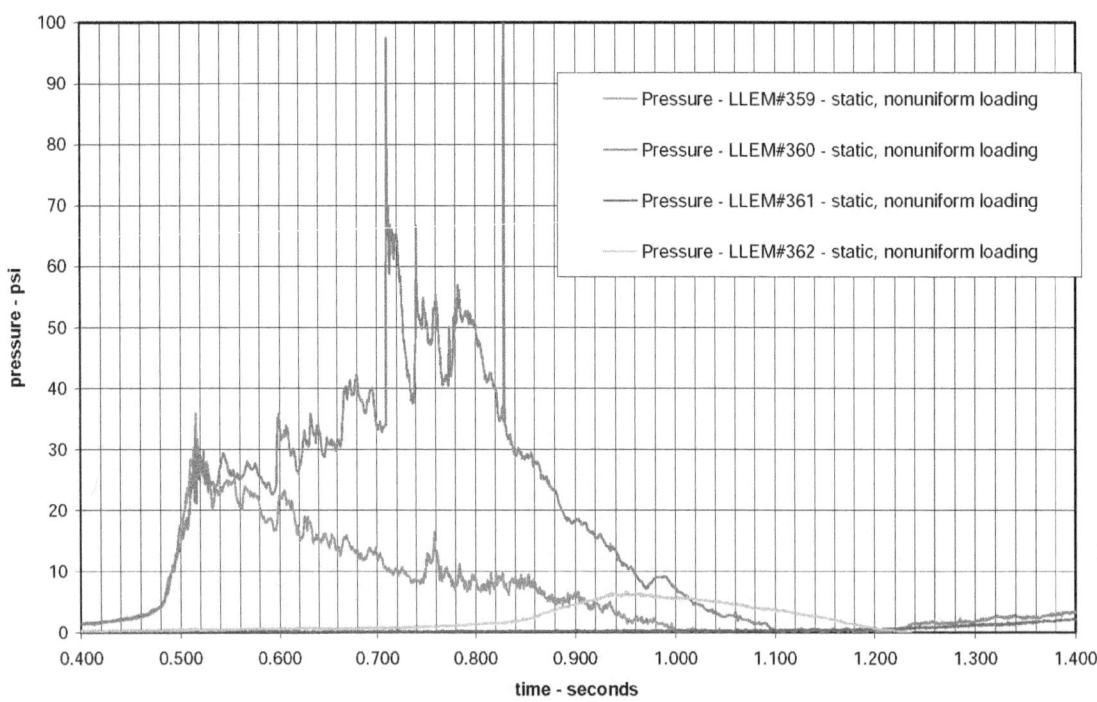

Figure A-10.—Category 1B - structure #1 - tests 9 to 12 - static, nonuniform loading. Meshblock seal - shotcrete with reinforcement - LLEM tests #359–362.

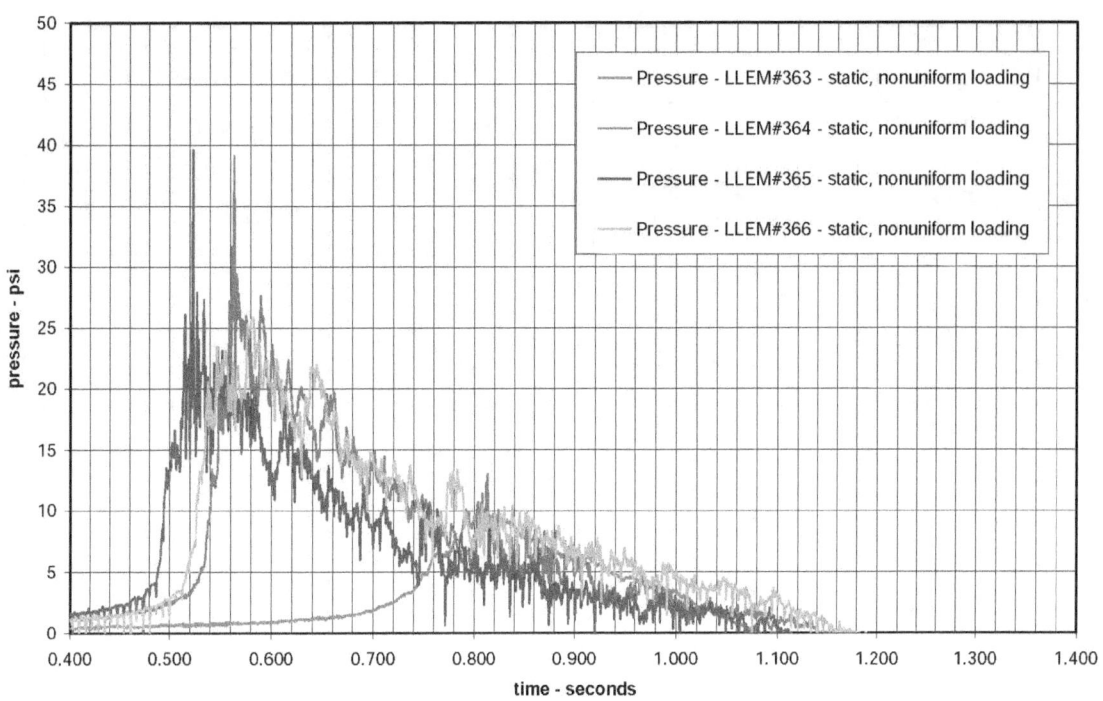

Figure A-11.—Category 1B - structure #1 - tests 13 to 16 - static, nonuniform loading. Meshblock seal - shotcrete with reinforcement - LLEM tests #363–366.

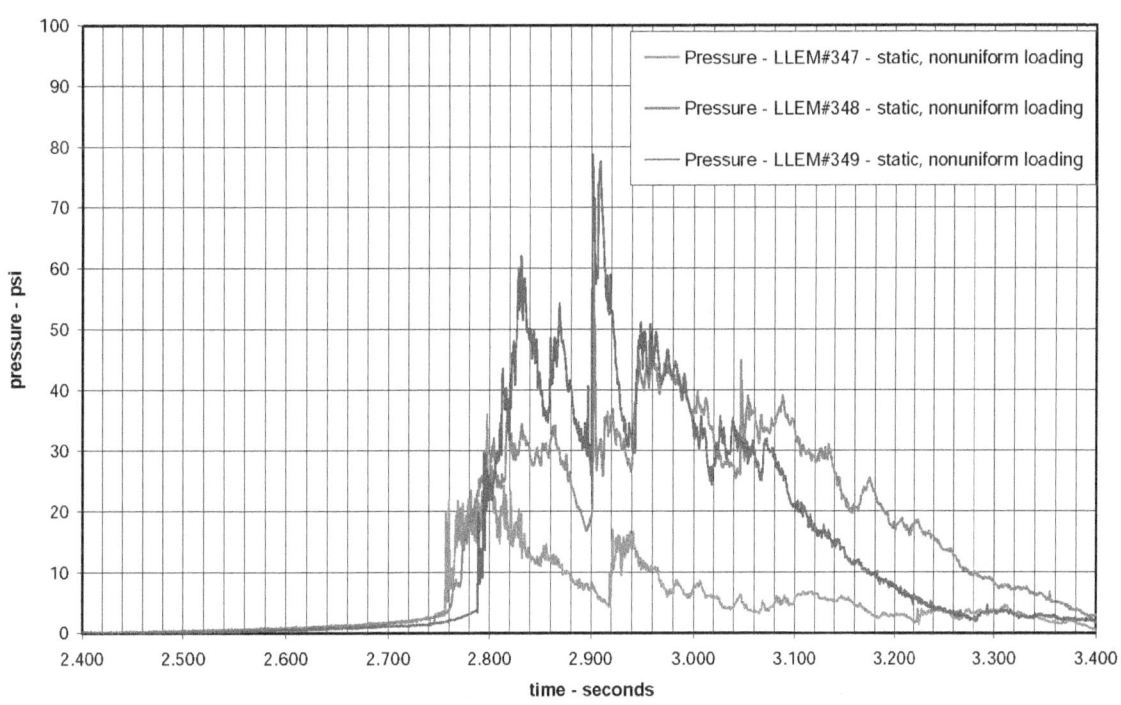

Figure A-12.—Category 1B - structure #2 - tests 1 to 3 - static, nonuniform loading. Meshblock seal - shotcrete with reinforcement - LLEM tests #347–349.

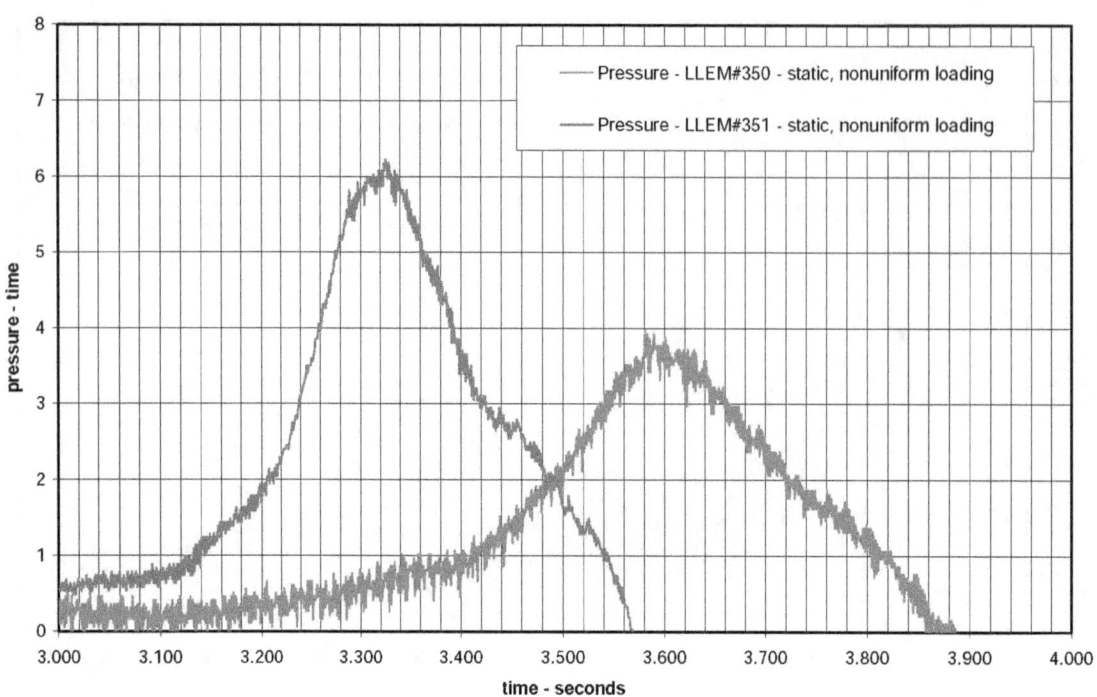

Figure A-13.—Category 1B - structure #2 - tests 4 and 5 - static, nonuniform loading. Meshblock seal - shotcrete with reinforcement - LLEM tests #350–351.

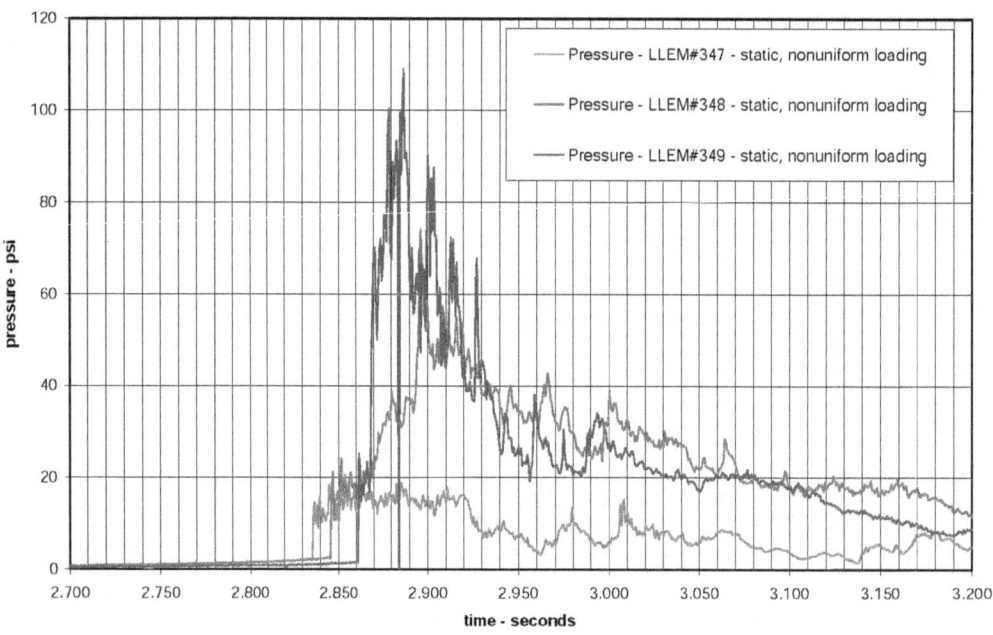

Figure A-14.—Category 1B - structure #3 - tests 1 to 3 - static, nonuniform loading. Meshblock seal - shotcrete with reinforcement - LLEM tests #347–349.

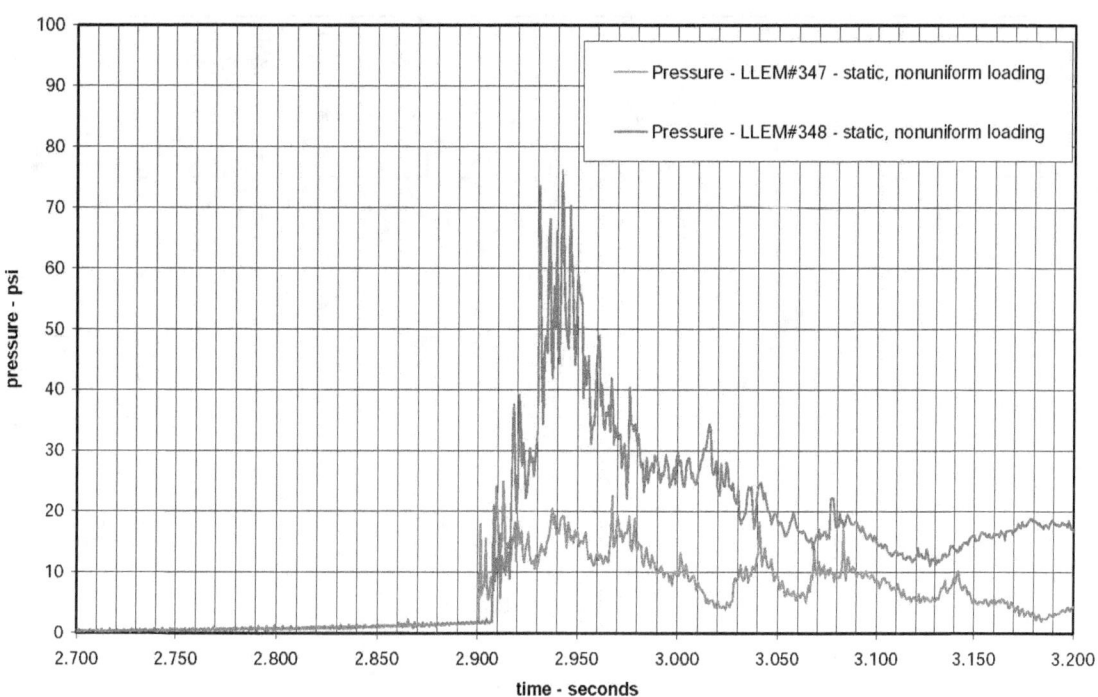

Figure A-15.—Category 1B - structure #4 - tests 1 and 2 - static, nonuniform loading.
Meshblock seal - shotcrete with reinforcement - LLEM tests #347–348.

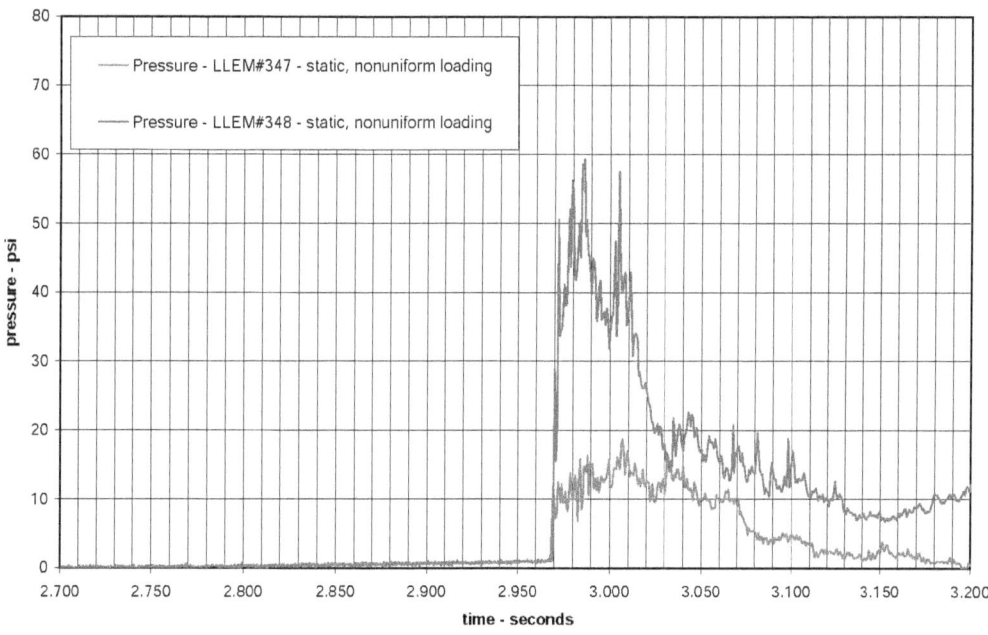

Figure A-16.—Category 1B - structure #5 - tests 1 and 2.
Meshblock seal - shotcrete with reinforcement - LLEM tests #347–348.

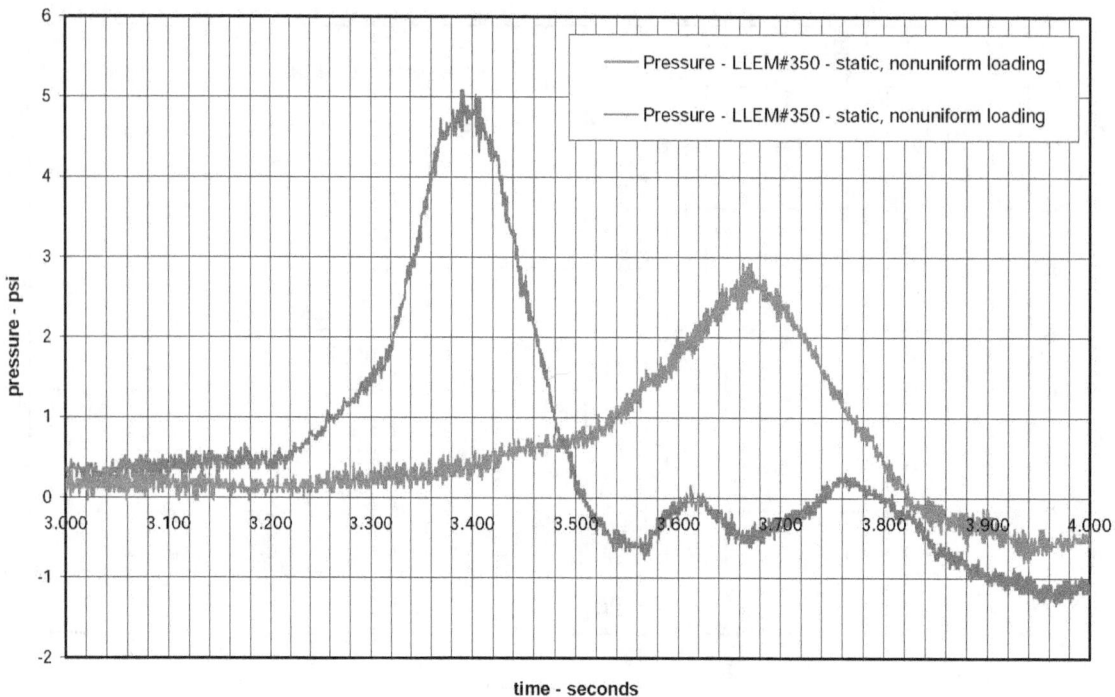

Figure A-17.—Category 1B - structure #6 - tests 1 and 2 - static, nonuniform loading. Meshblock seal - shotcrete with reinforcement - LLEM tests #350–351.

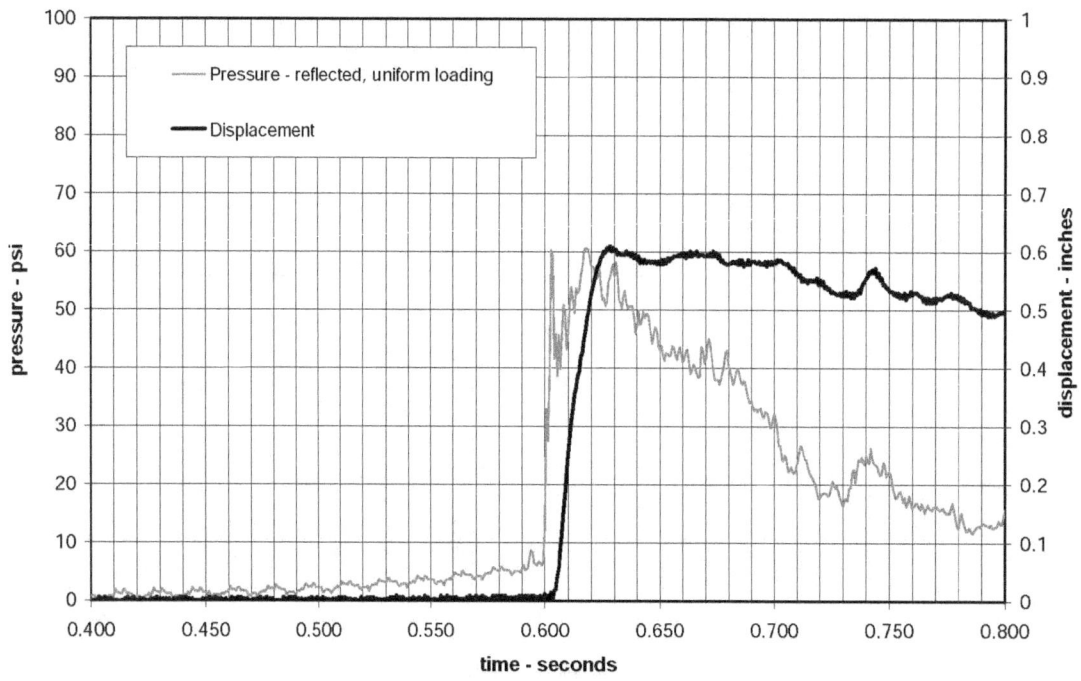

Figure A-18.—Category 2A - structure #1 - test 1 - reflected, uniform loading. Pumpable 48 in - LLEM test #508.

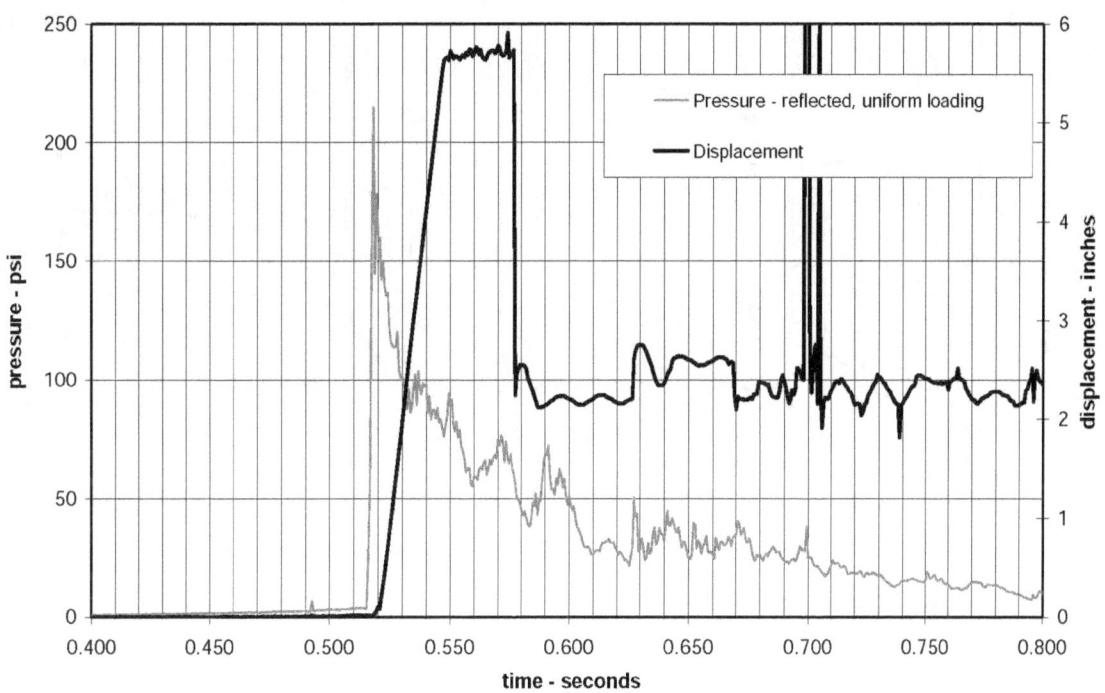

Figure A-19.—Category 2A - structure #1 - test 2 - reflected, uniform loading. Pumpable 48 in - LLEM test #509.

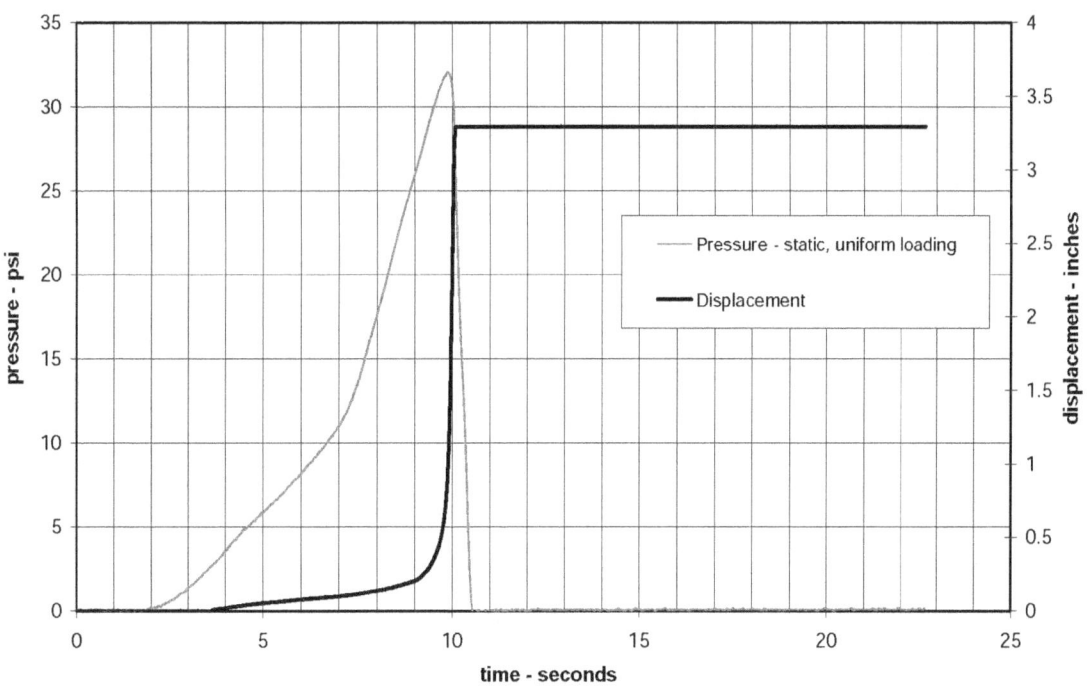

Figure A-20.—Category 2A - structure #2 - test 1 - static, uniform loading. Pumpable 48 in - test C3-44E.

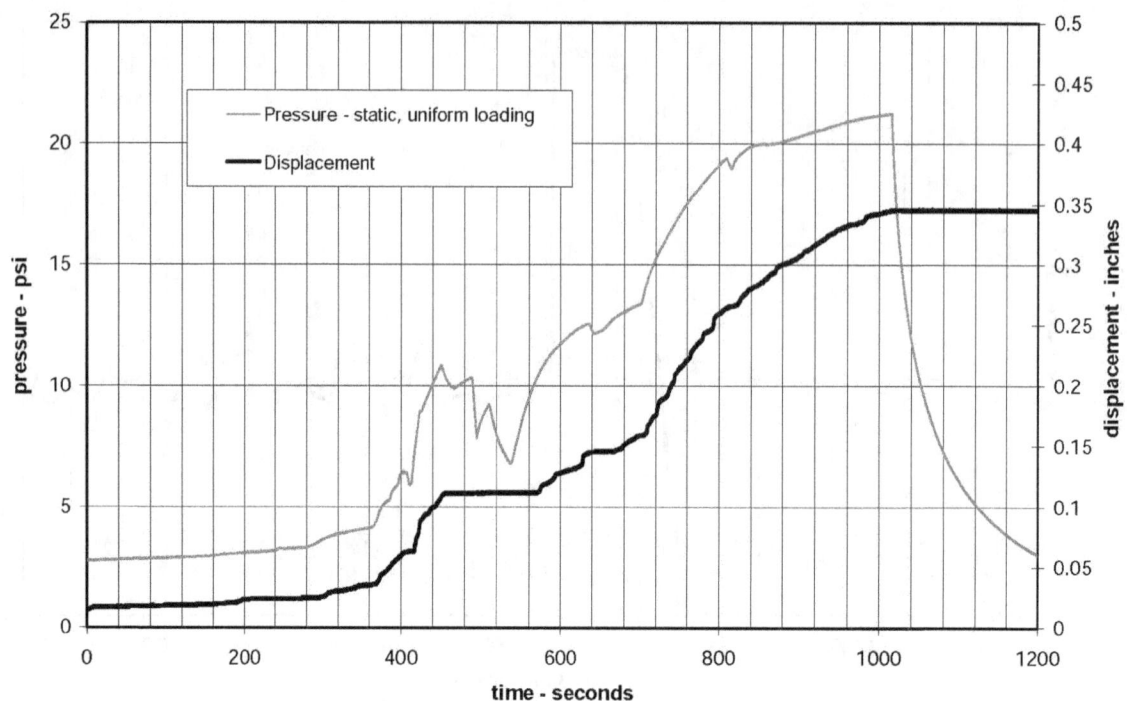

Figure A-21.—Category 2A - structure #3 - test 1 - static, uniform loading.
Pumpable 48 in - test C7-64W.

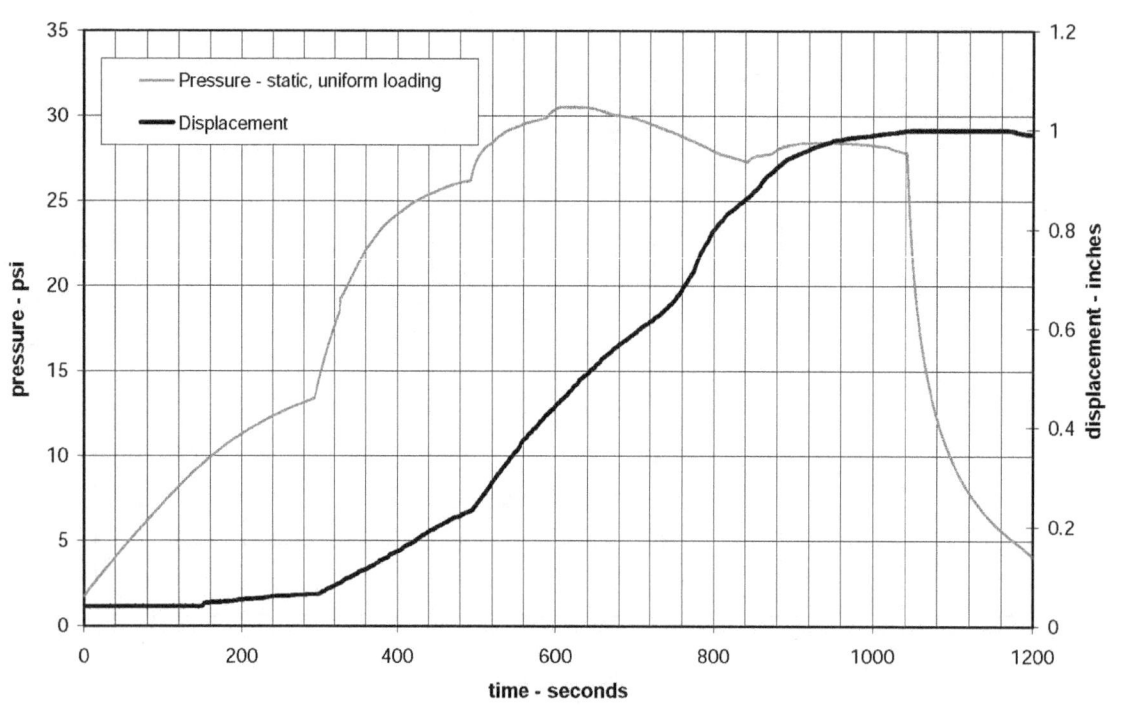

Figure A-22.—Category 2A - structure #3 - test 2 - static, uniform loading.
Pumpable 48 in - test C7-68W.

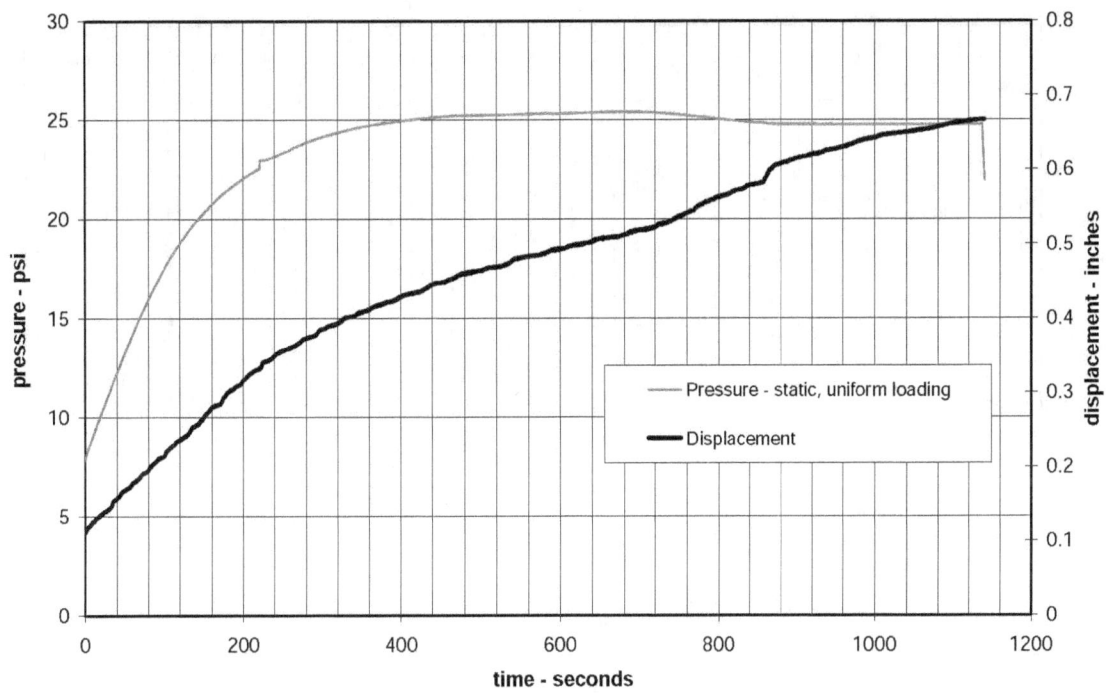

Figure A-23.—Category 2A - structure #3 - test 3 - static, uniform loading. Pumpable 48 in - test C7-70W.

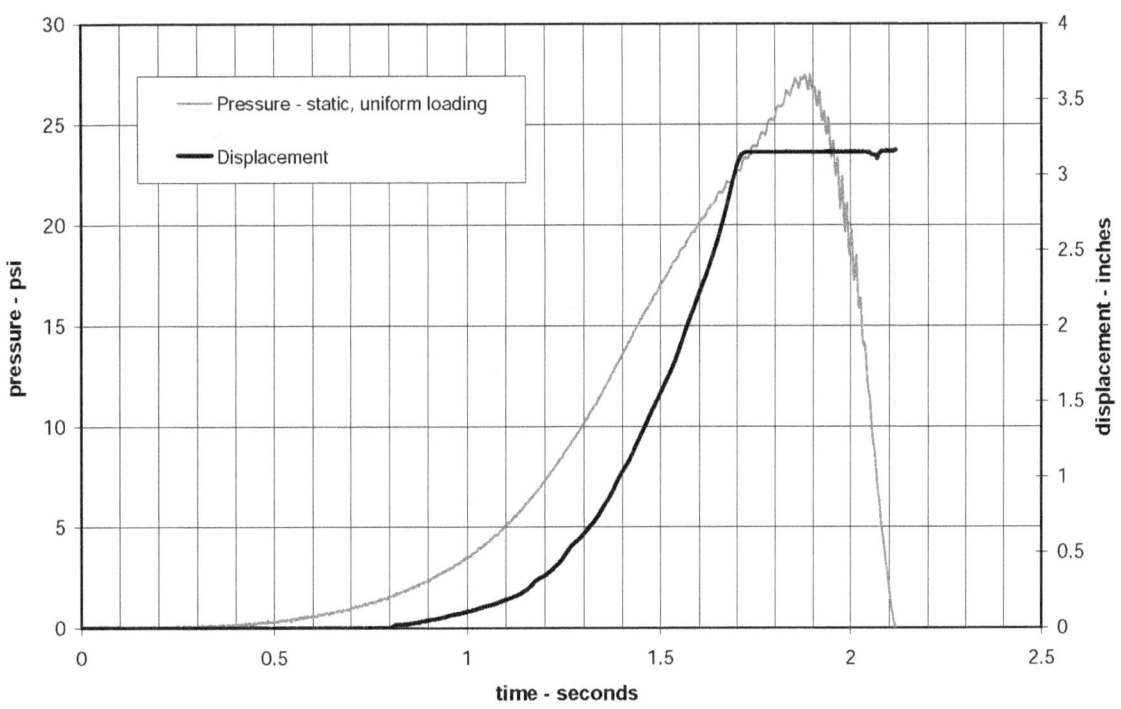

Figure A-24.—Category 2A - structure # 4 - test 1 - static, uniform loading. Pumpable 48 in - test L2-51E.

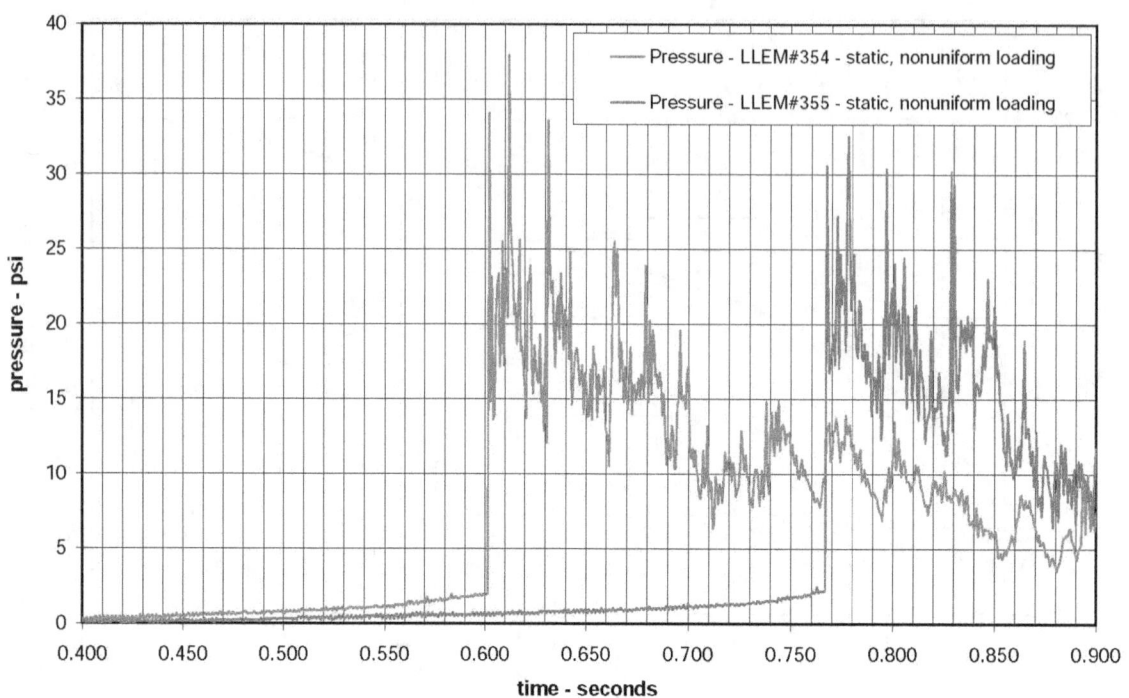

Figure A-25.—Category 2B - structure #1 - tests 1 and 2 - static, nonuniform loading. Pumpable 36 in - LLEM tests #354–355.

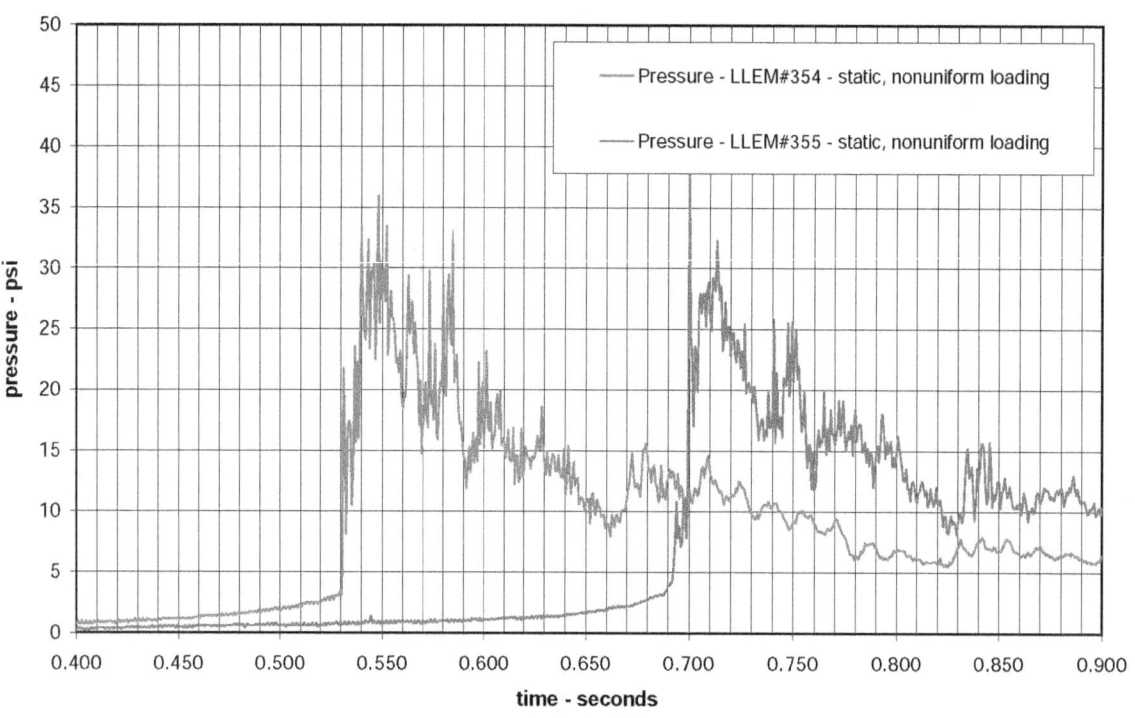

Figure A-26.—Category 2C - structure #1 - tests 1 and 2 - static, nonuniform loading. Pumpable 24 in - LLEM tests #354–355.

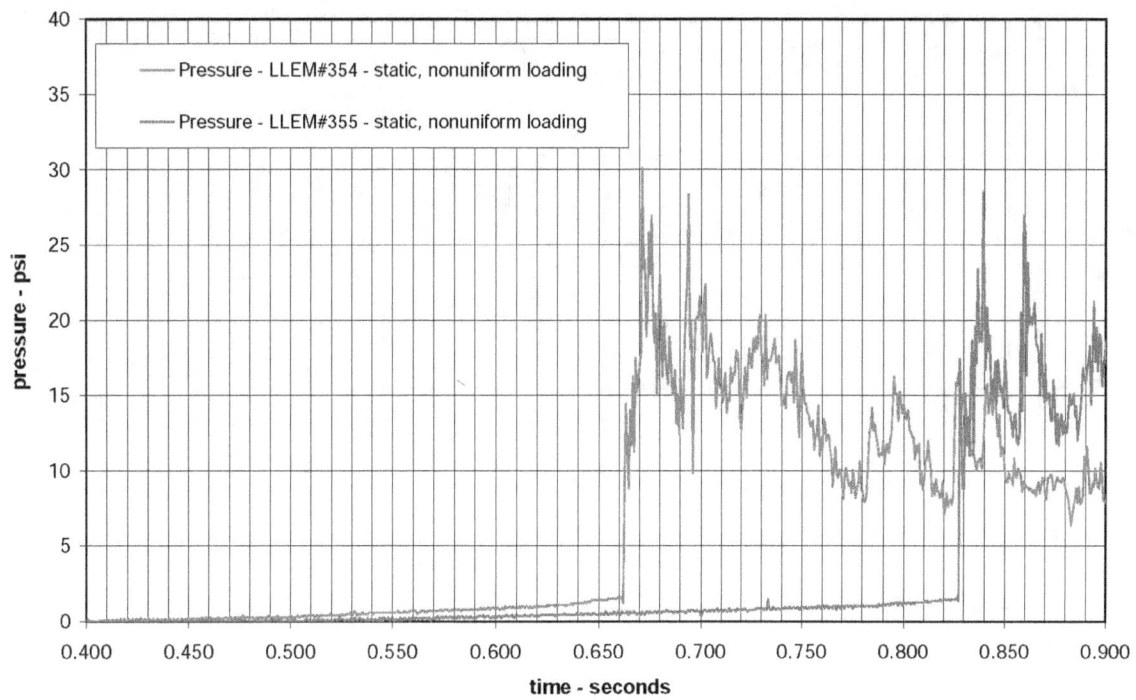

Figure A-27.—Category 2C - structure #2 - tests 1 and 2 - static, nonuniform loading.
Pumpable 24 in - LLEM tests #354–355.

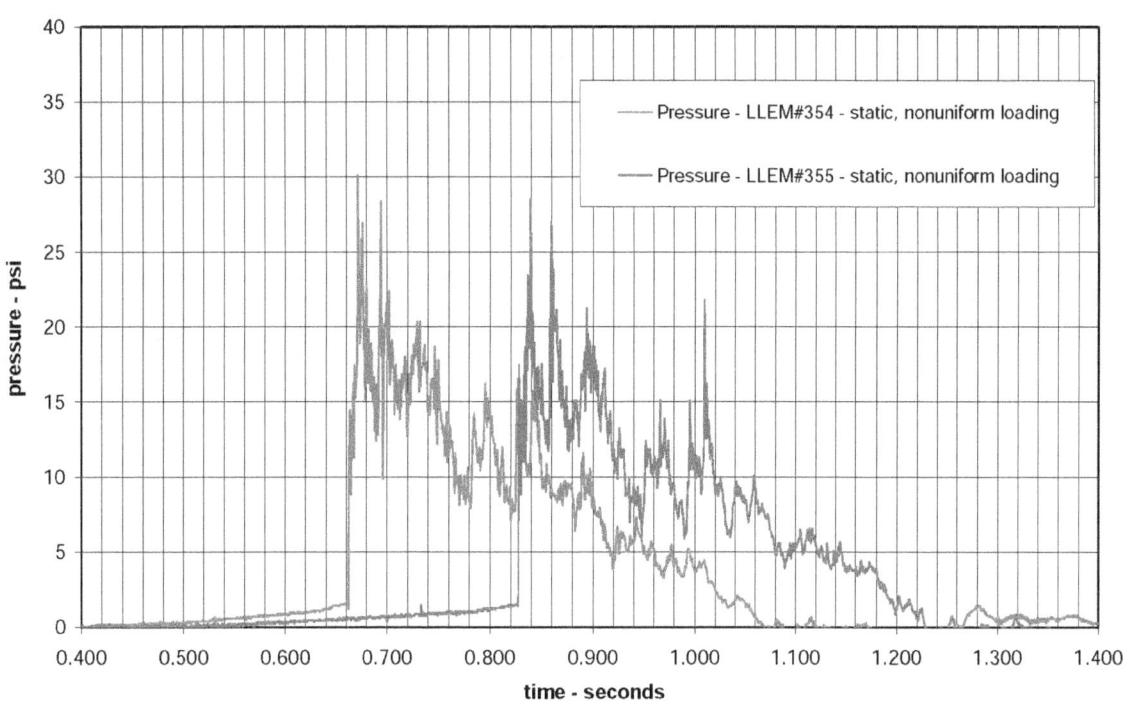

Figure A-28.—Category 2C - structure #3 - tests 1 and 2 - static, nonuniform loading.
Pumpable 24 in - LLEM tests #354–355.

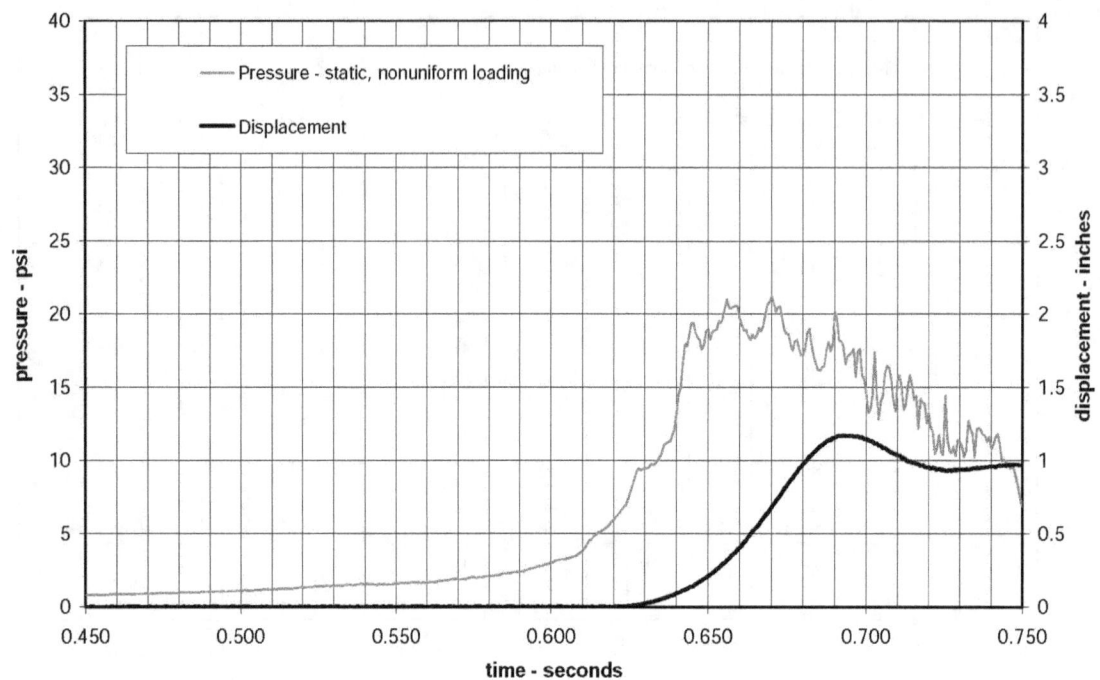

Figure A-29.—Category 2C - structure #4 - test 1 - static, nonuniform loading.
Pumpable 24 in - LLEM test #403.

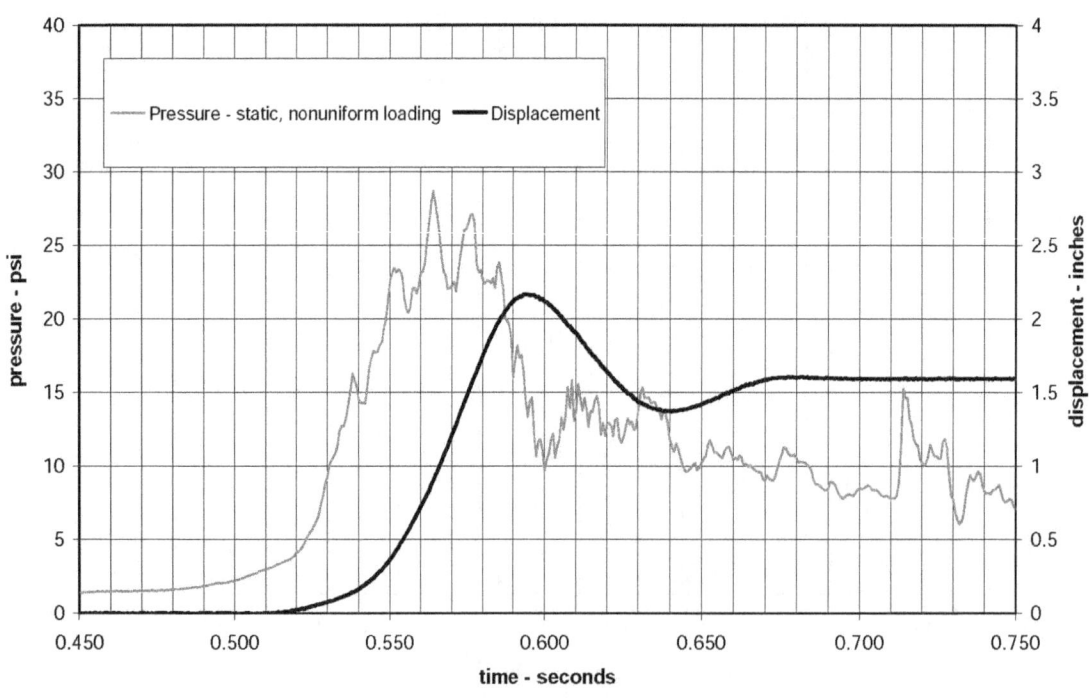

Figure A-30.—Category 2C - structure #4 - test 2 - static, nonuniform loading.
Pumpable 24 in - LLEM test #404.

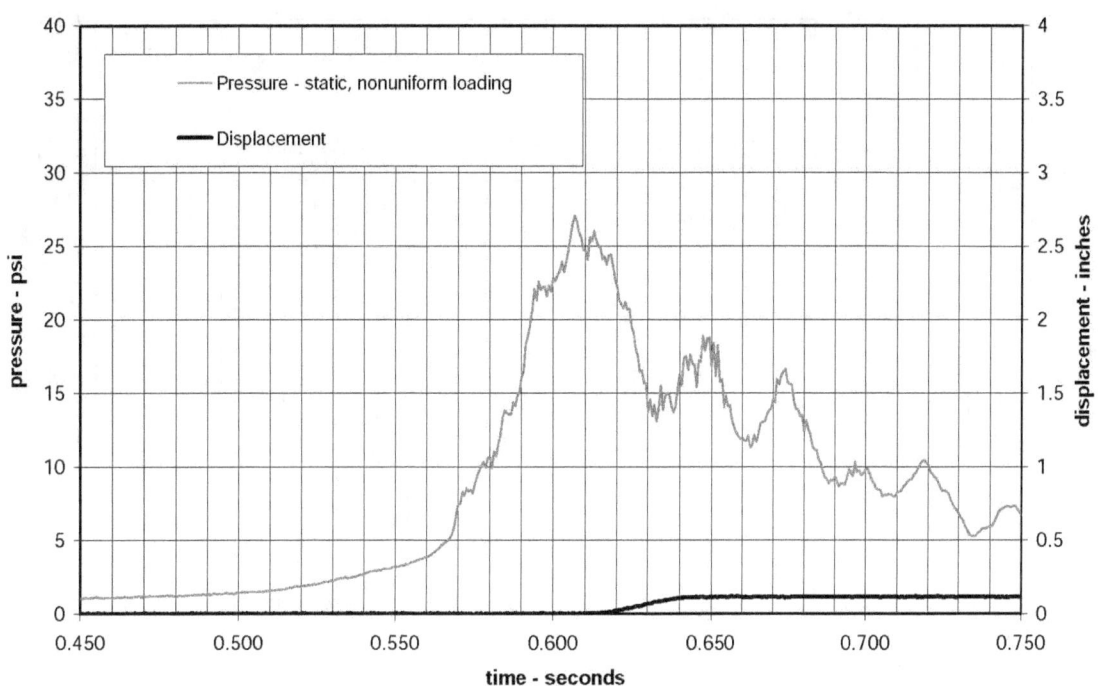

Figure A-31.—Category 2C - structure #4 - test 3 - static, nonuniform loading. Pumpable 24 in - LLEM test #405.

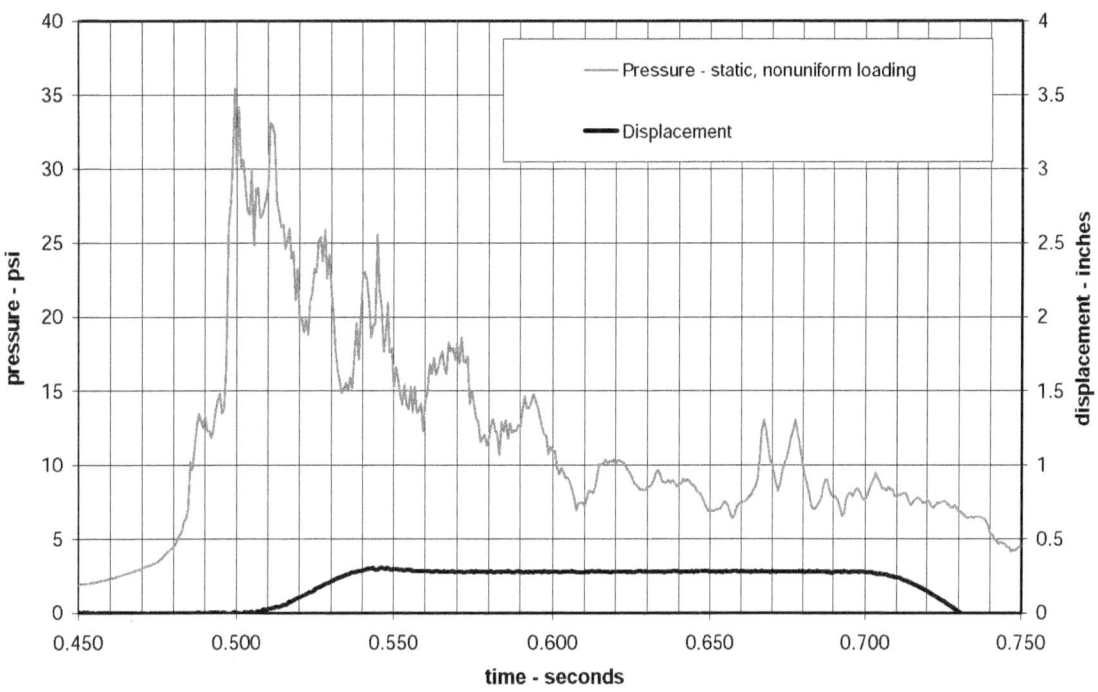

Figure A-32.—Category 2C - structure #4 - test 4 - static, nonuniform loading. Pumpable 24 in - LLEM test #406.

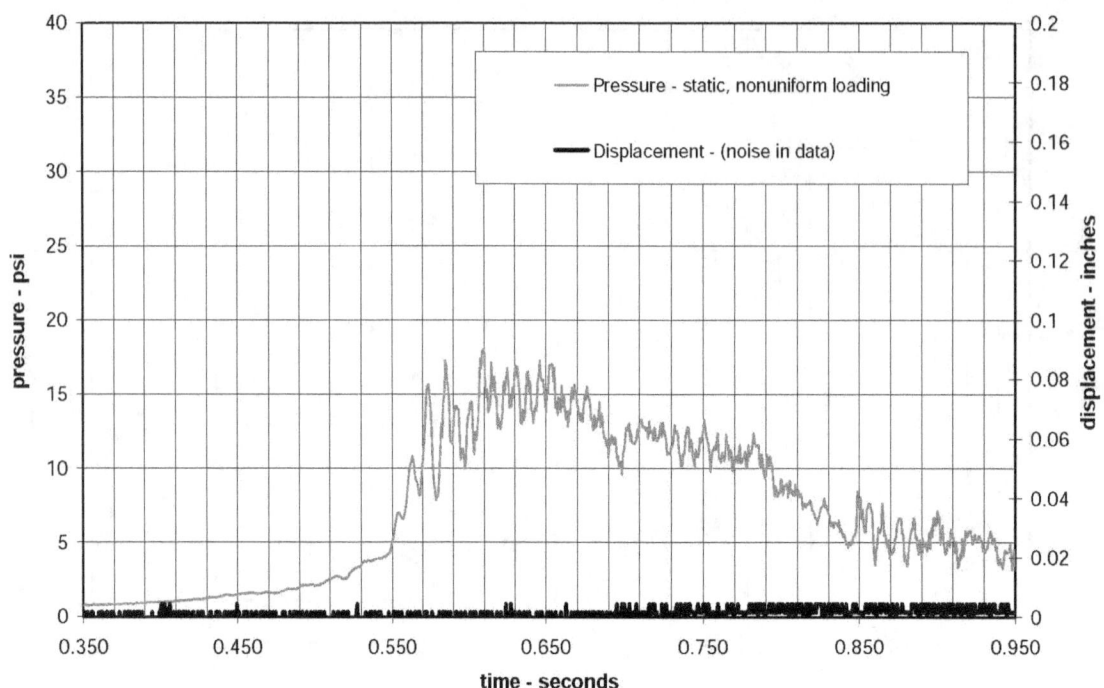

Figure A-33.—Category 3A - structure #1 - test 1 - static, nonuniform loading.
Standard solid-concrete-block seal - LLEM test #403.

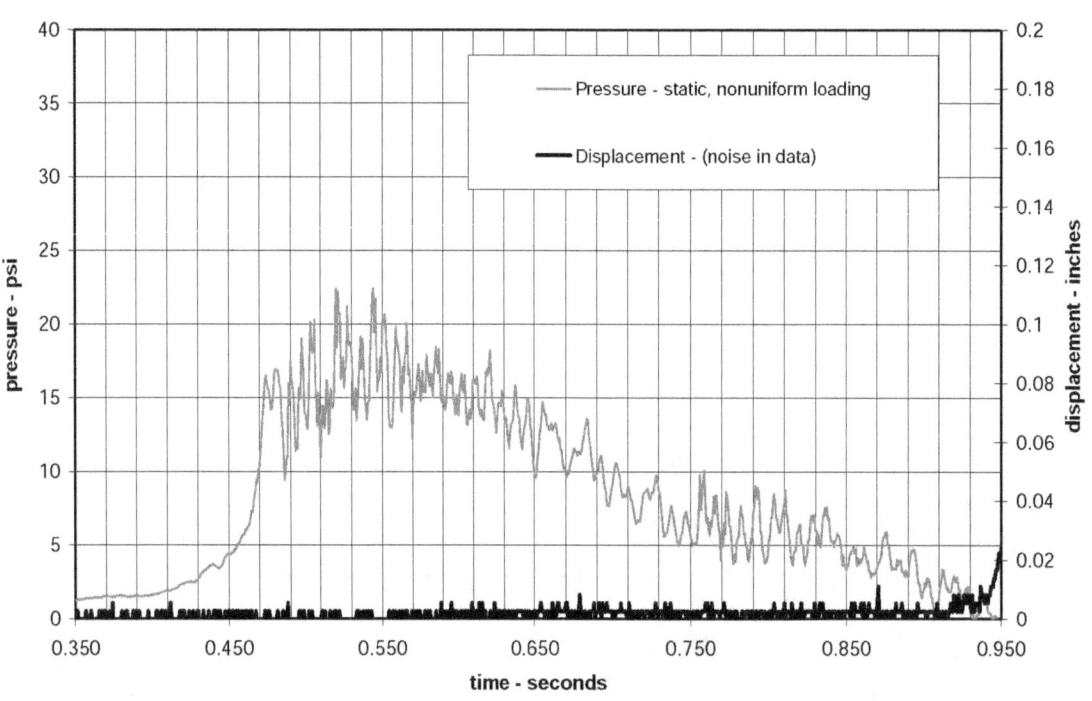

Figure A-34.—Category 3A - structure #1 - test 2 - static, nonuniform loading.
Standard solid-concrete-block seal - LLEM test #404.

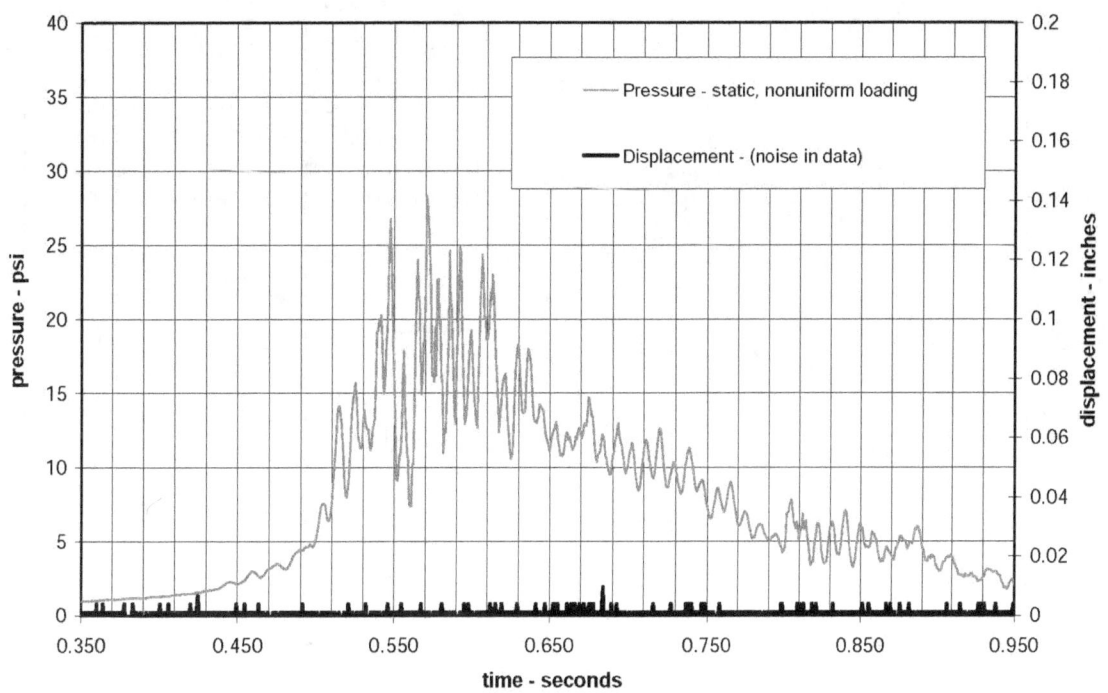

Figure A-35.—Category 3A - structure #1 - test 3 - static, nonuniform loading. Standard solid-concrete-block seal - LLEM test #405.

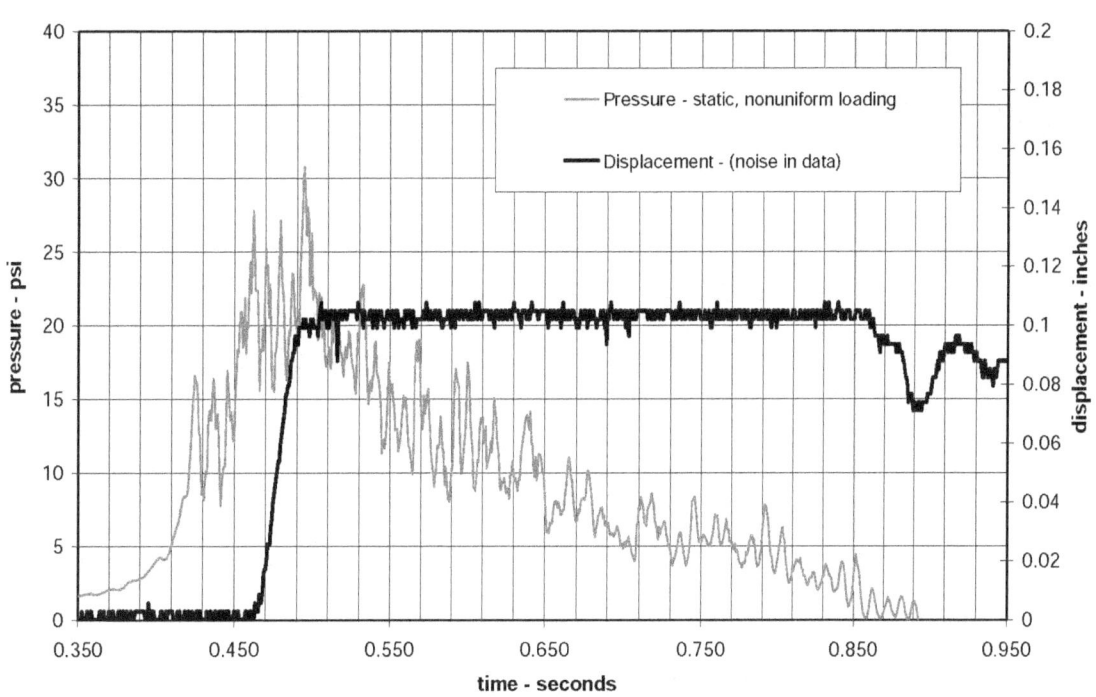

Figure A-36.—Category 3A - structure #1 - test 4 - static, nonuniform loading. Standard solid-concrete-block seal - LLEM test #406.

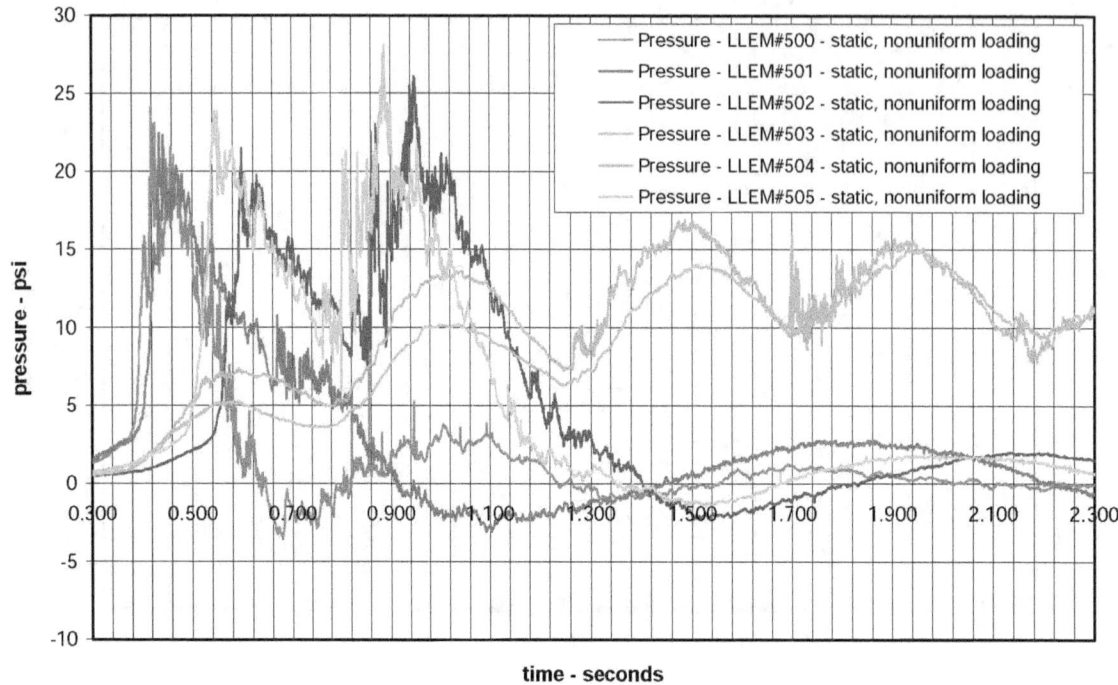

Figure A-37.—Category 3A - structure #2 - tests 1 to 6 - static, nonuniform loading. Standard solid-concrete-block seal - LLEM tests #500–505.

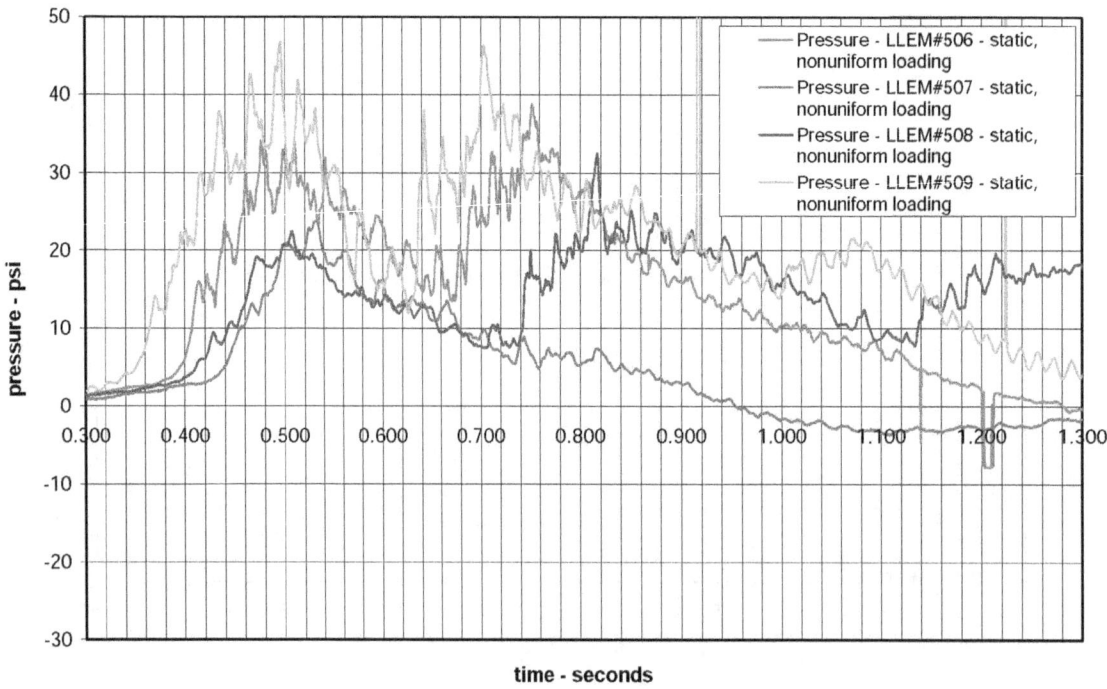

Figure A-38.—Category 3A - structure #2 - tests 7 to 10 - static, nonuniform loading. Standard solid-concrete-block seal - LLEM tests #506–509.

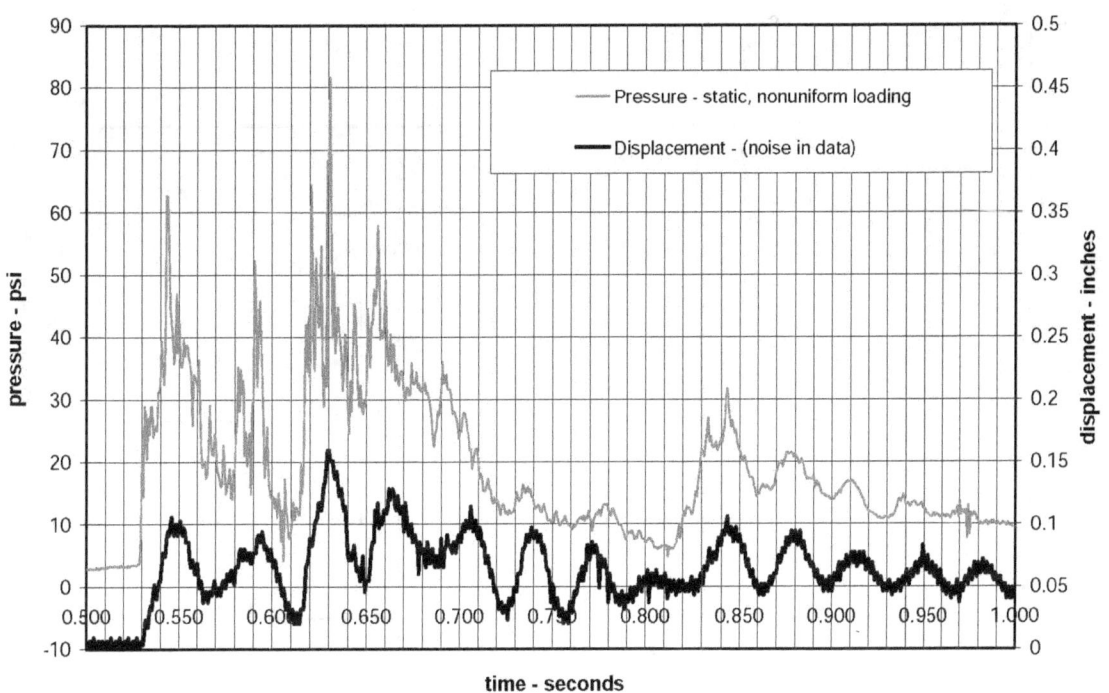

Figure A-39.—Category 3A - structure #3 - test 1 - static, nonuniform loading.
Standard solid-concrete-block seal - LLEM test #506.

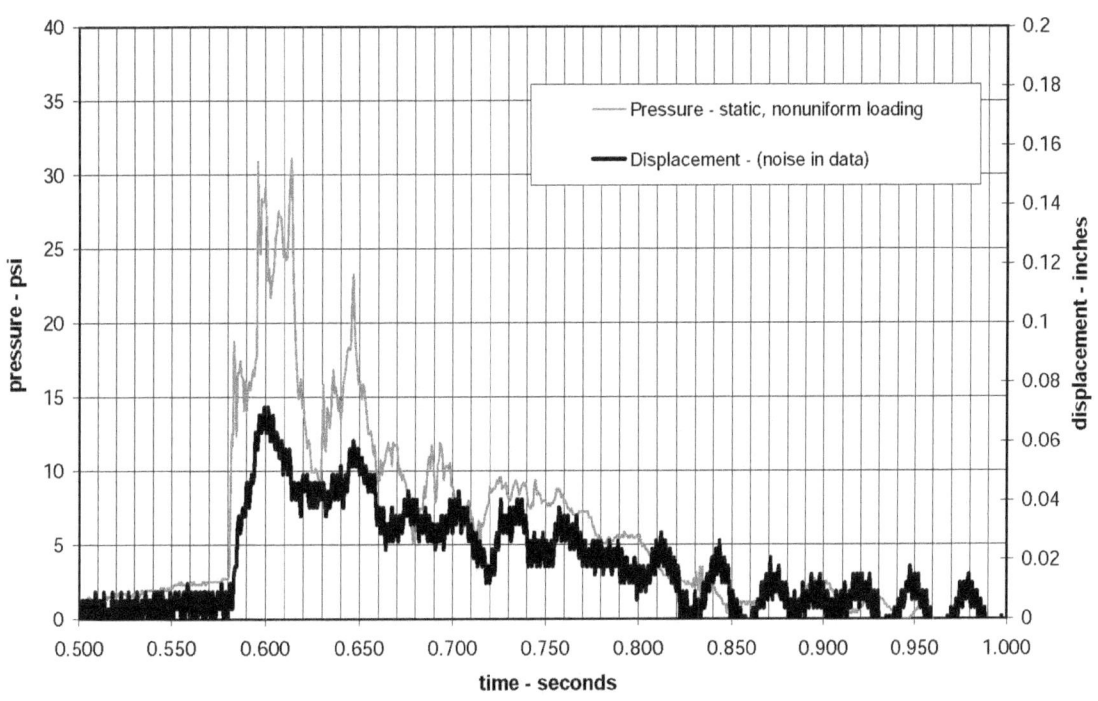

Figure A-40.—Category 3A - structure #3 - test 2 - static, nonuniform loading.
Standard solid-concrete-block seal - LLEM test #507.

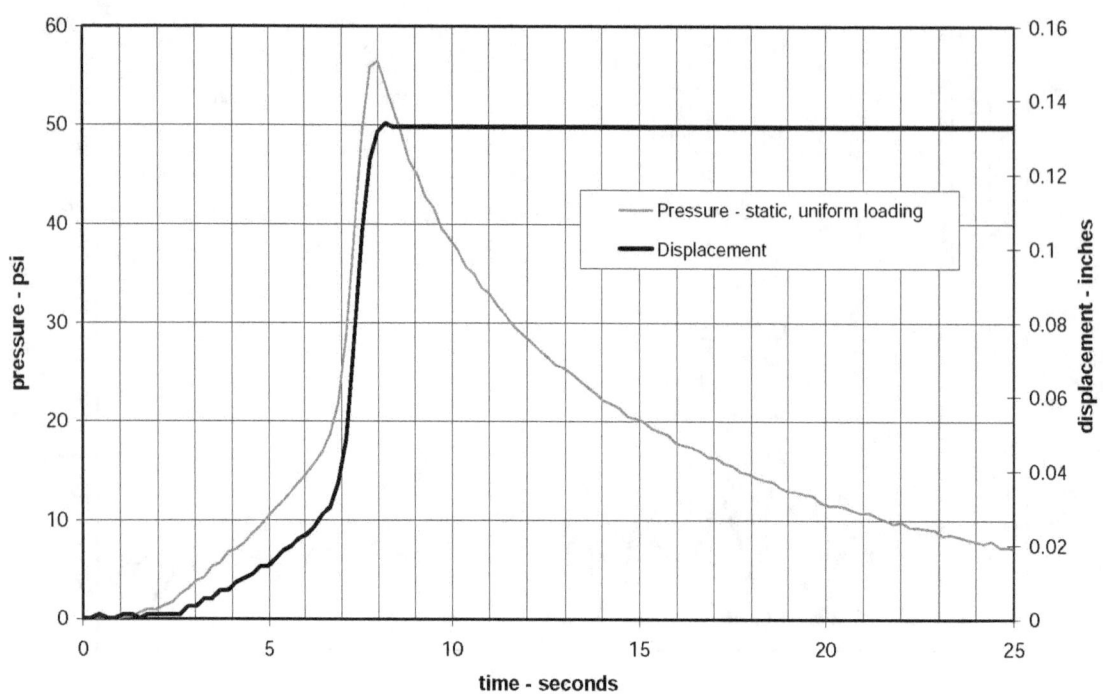

Figure A-41.—Category 3A - structure #4 - test 1 - static, uniform loading. Standard solid-concrete-block seal - test C1-5E.

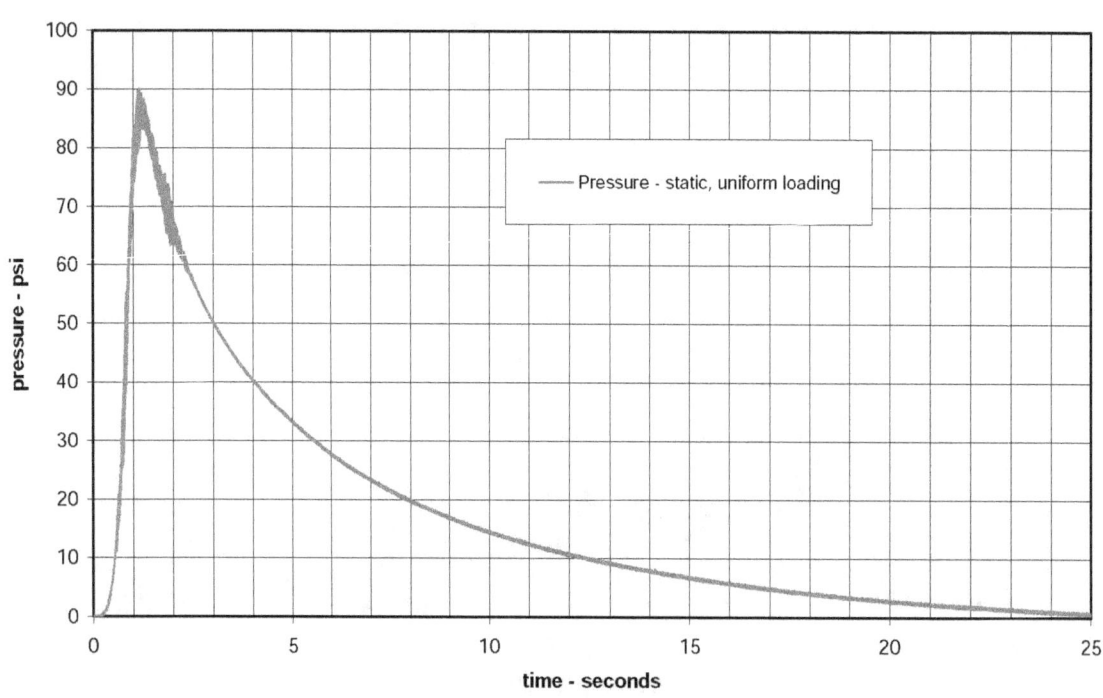

Figure A-42.—Category 3A - structure #4 - test 2 - static, uniform loading. Standard solid-concrete-block seal - test C1-8E.

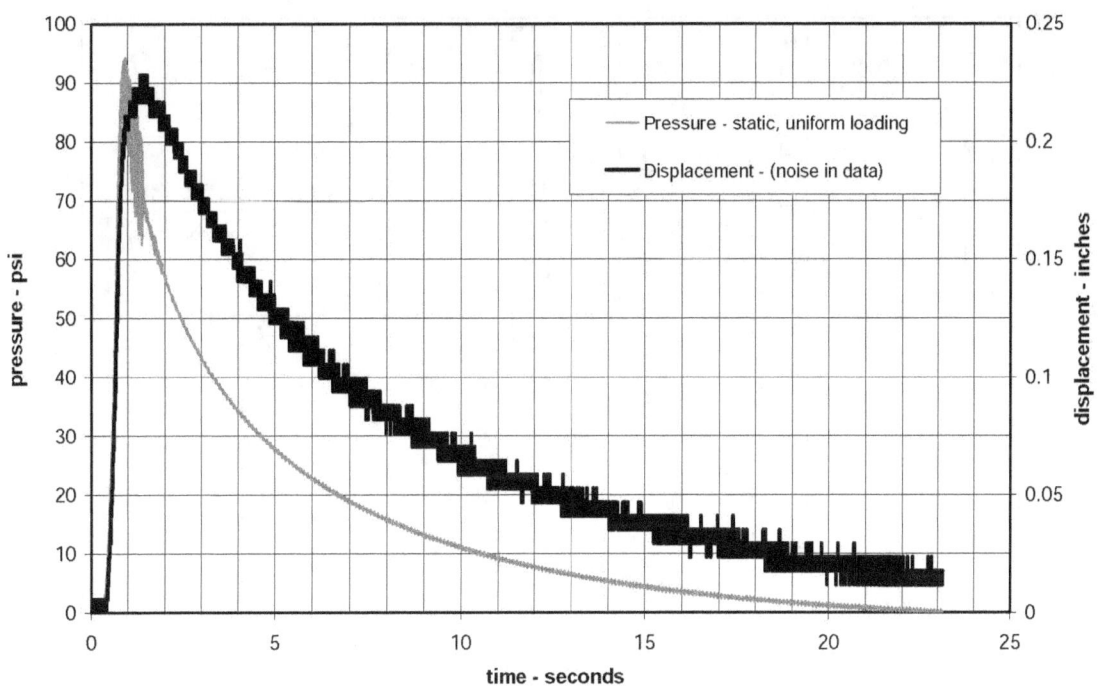

Figure A-43.—Category 3A - structure #4 - test 3 - static, uniform loading.
Standard solid-concrete-block seal - test C1-9E.

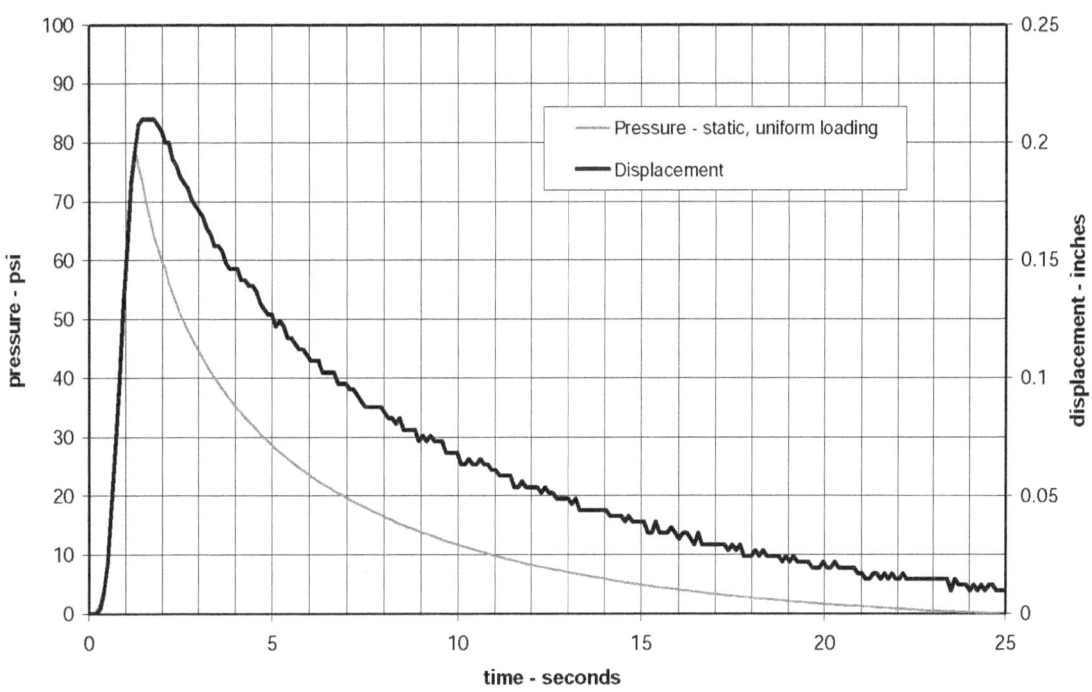

Figure A-44.—Category 3A - structure #4 - test 4 - static, uniform loading.
Standard solid-concrete-block seal - test C1-10E.

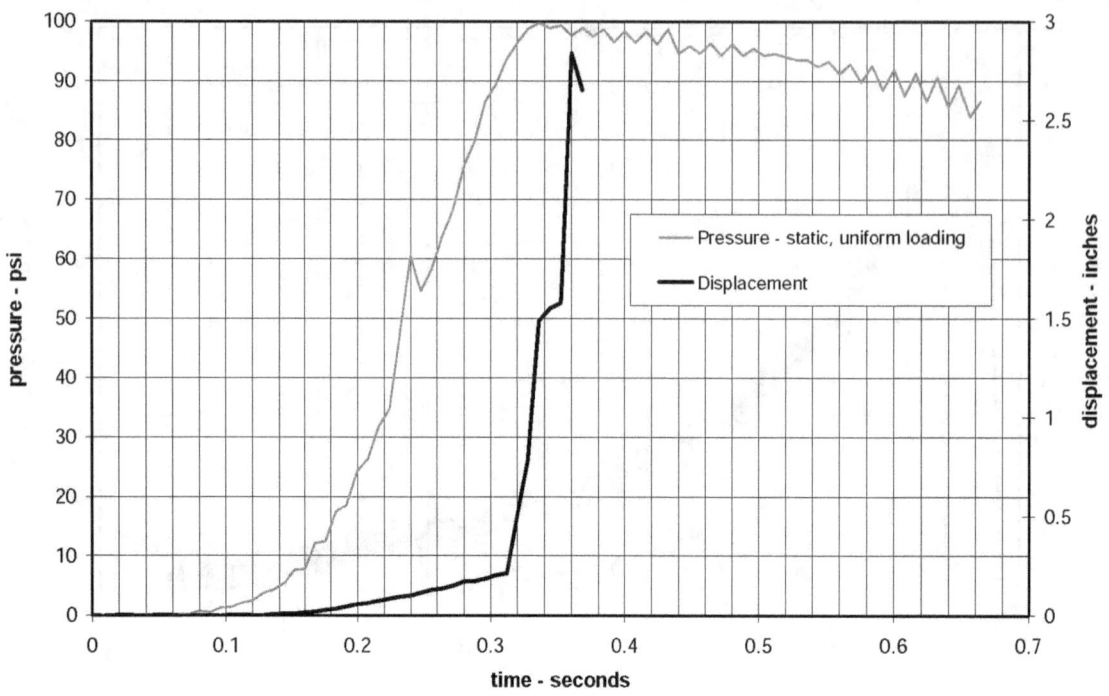

Figure A-45.—Category 3A - structure #4 - test 5 - static, uniform loading. Standard solid-concrete-block seal - test C1-11E.

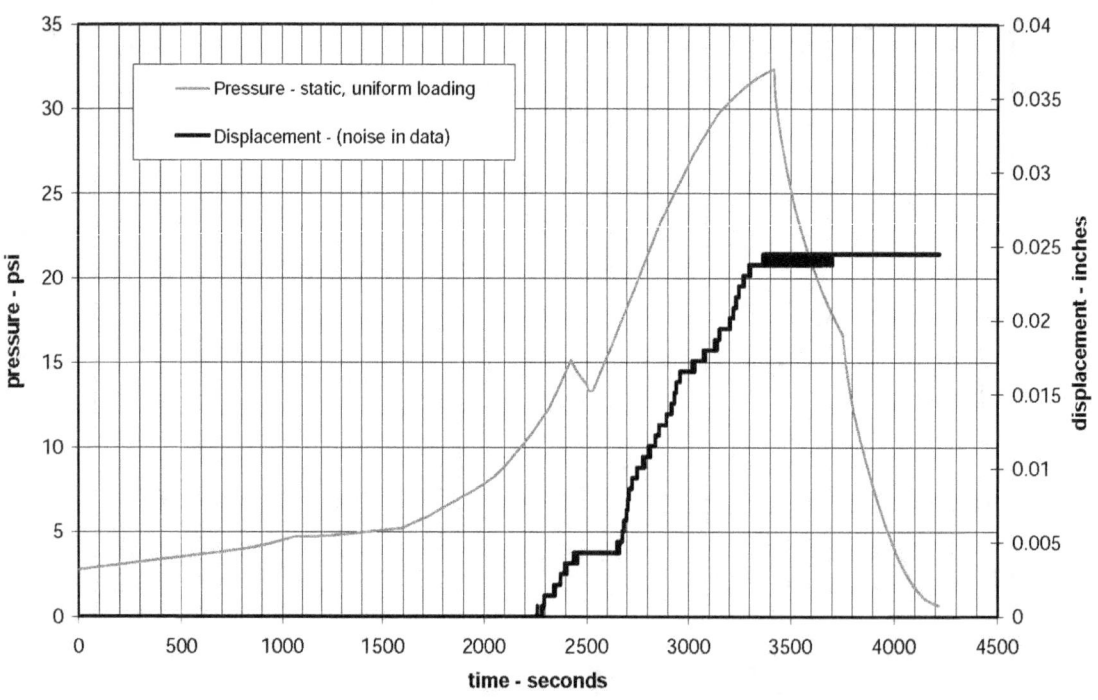

Figure A-46.—Category 3A - structure #5 - test 1 - static, uniform loading. Standard solid-concrete-block seal - test C6-60W.

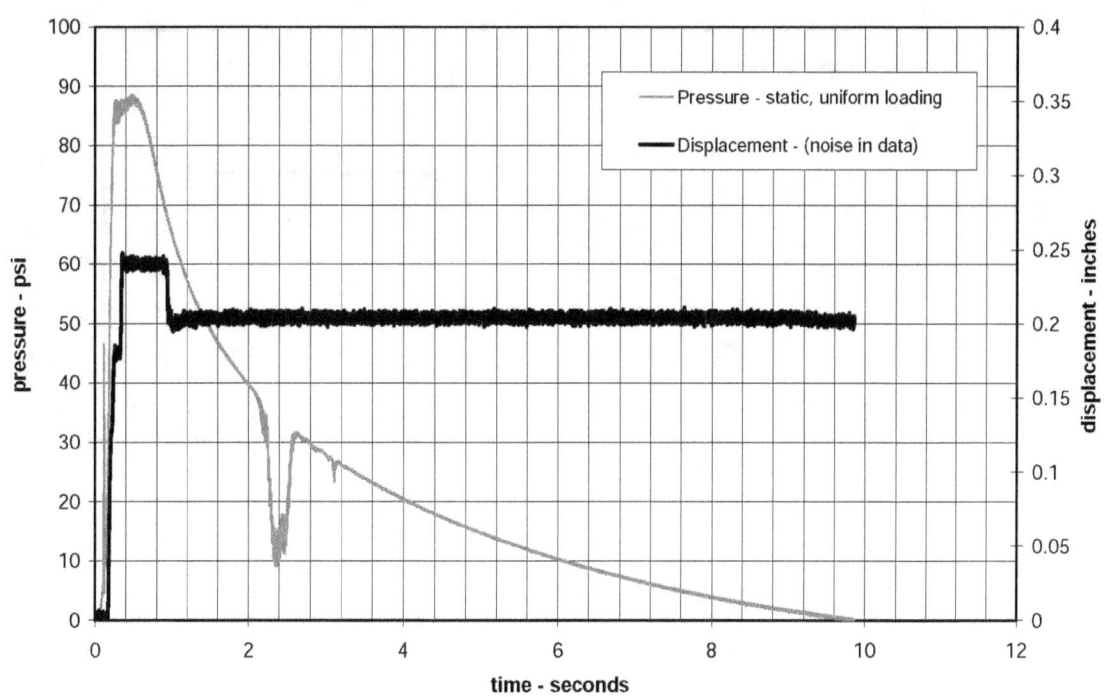

Figure A-47.—Category 3A - structure #5 - test 2 - static, uniform loading. Standard solid-concrete-block seal - test C6-62E.

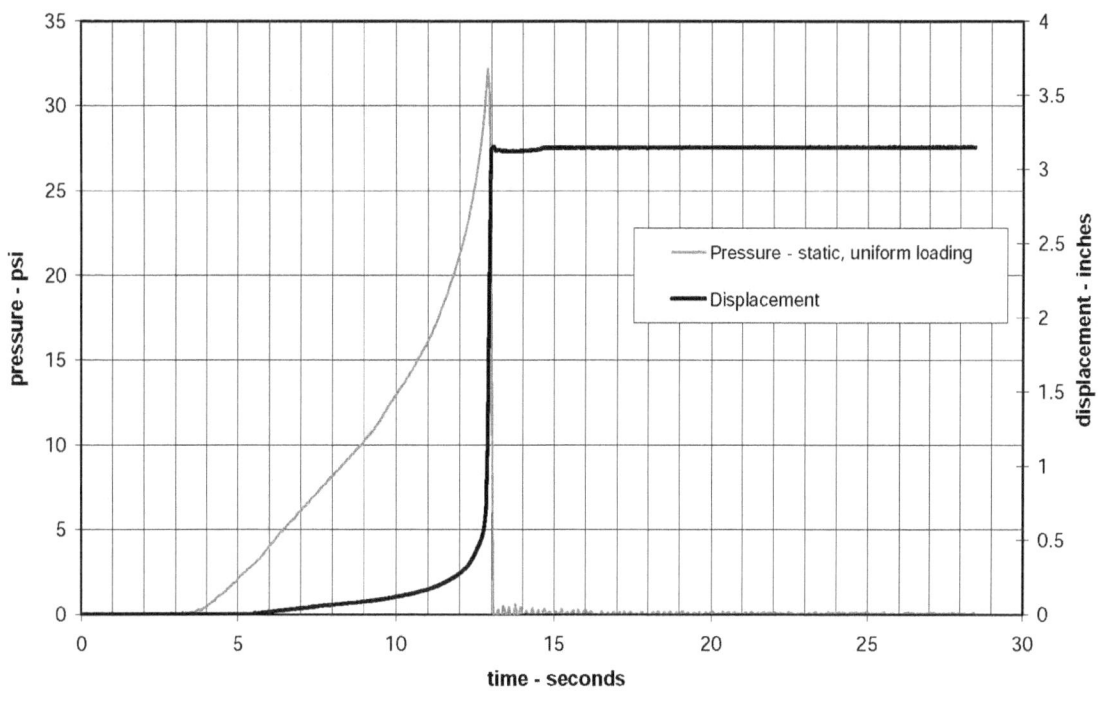

Figure A-48.—Category 3A - structure #6 - test 1 - static, uniform loading. Standard solid-concrete-block seal - test L1-37E.

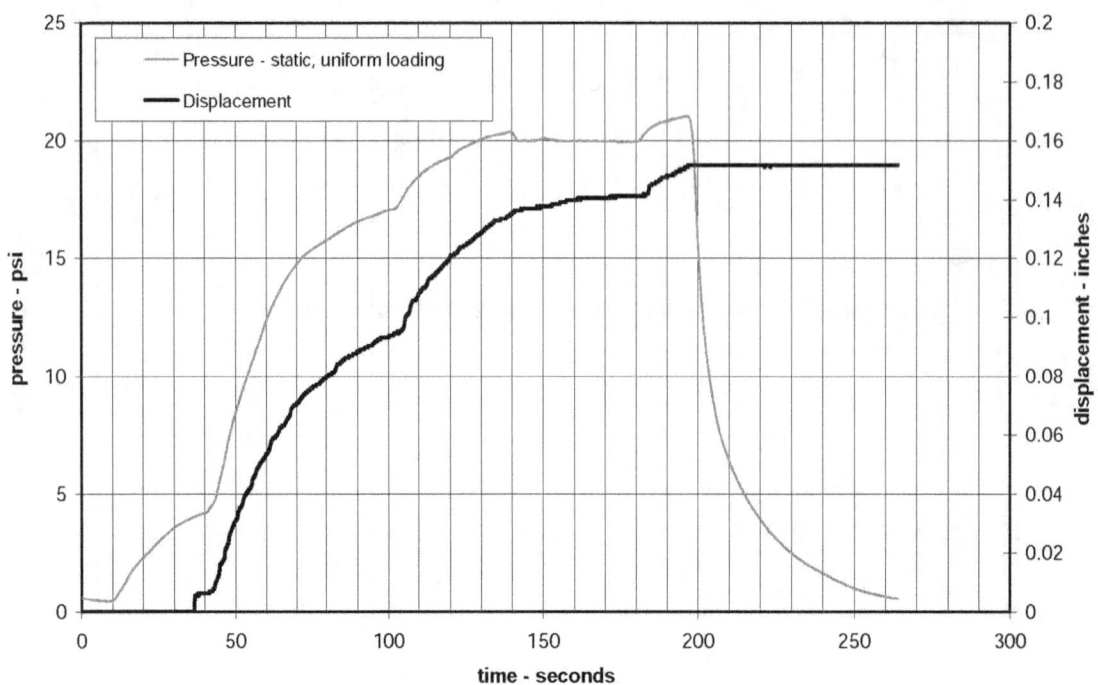

Figure A-49.—Category 3A - structure #7 - test 1 - static, uniform loading. Standard solid-concrete-block seal - test SRCM 1.

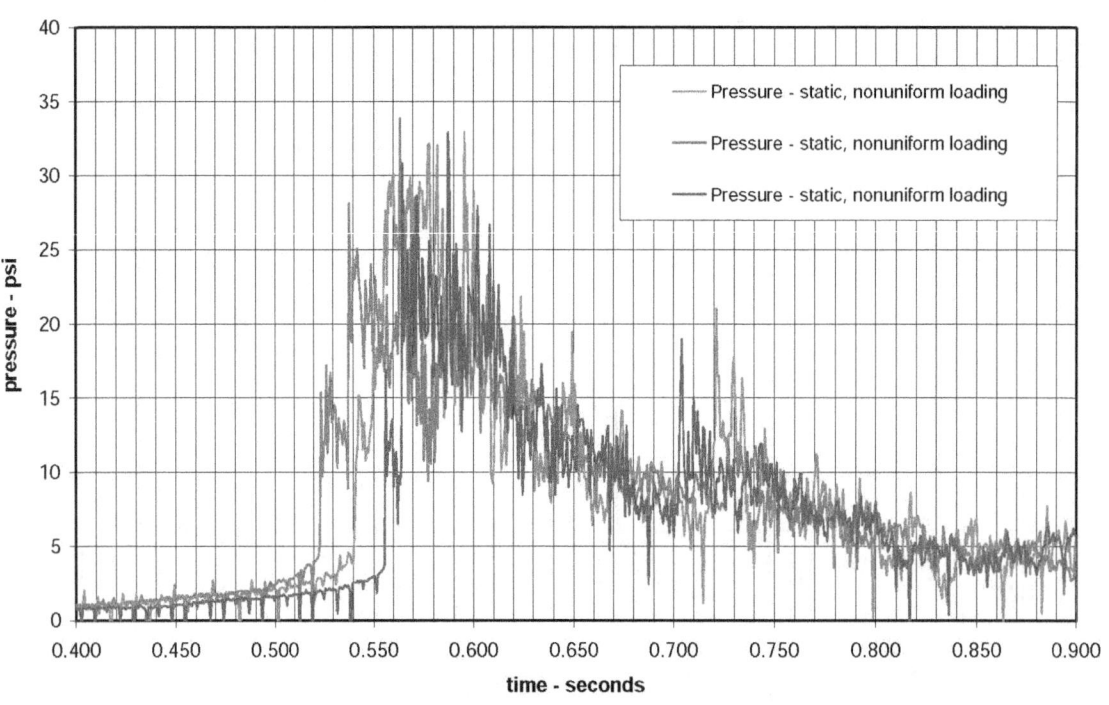

Figure A-50.—Category 3B - structure #1 - test 1 - static, nonuniform loading. Solid-concrete-block seal with Packsetter bags - LLEM test #365.

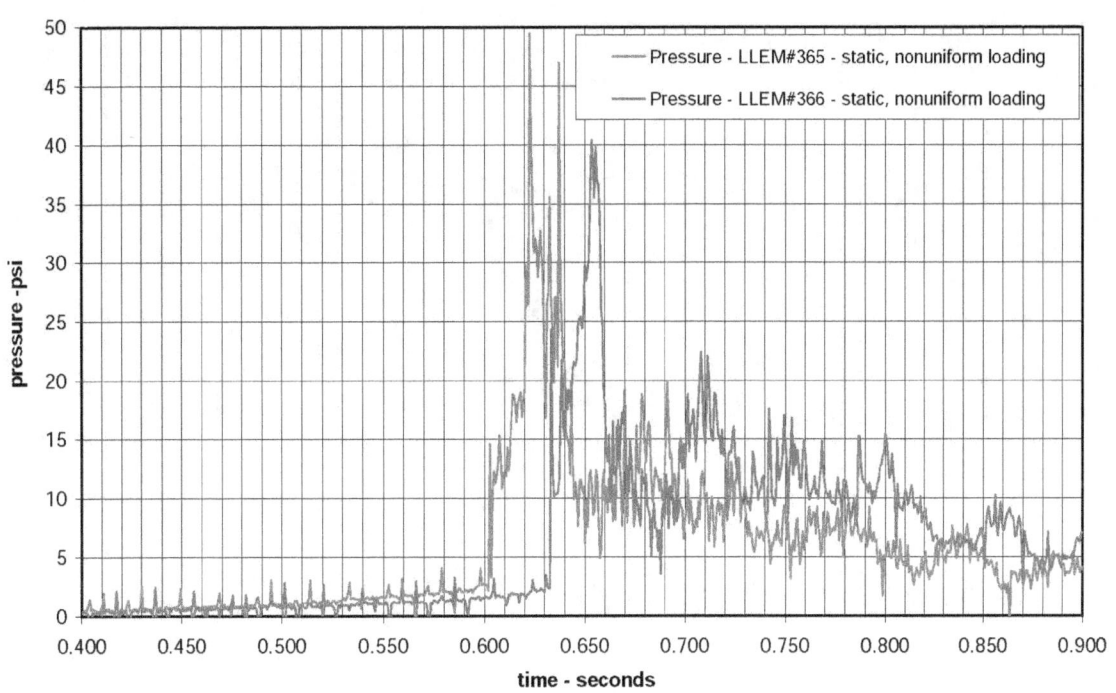

Figure A-51.—Category 3B - structure #2 - tests 1 and 2 - static, nonuniform loading. Solid-concrete-block seal with Packsetter bags - LLEM tests #365–366.

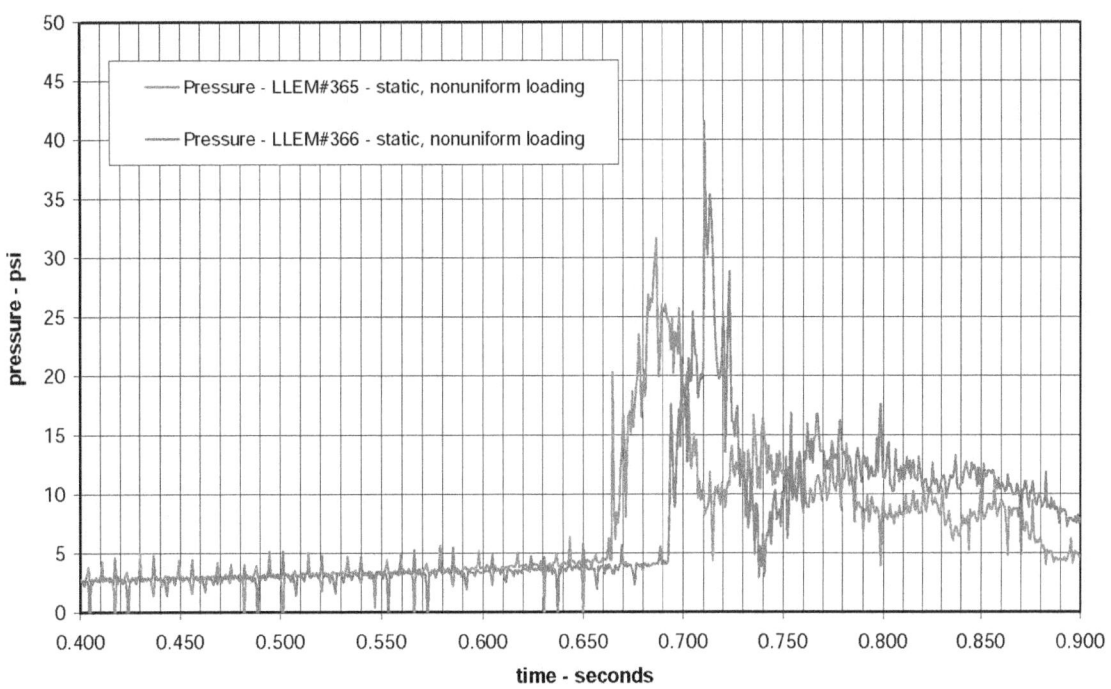

Figure A-52.—Category 3B - structure #3 - tests 1 and 2 - static, nonuniform loading. Solid-concrete-block seal with Packsetter bags - LLEM tests #365–366.

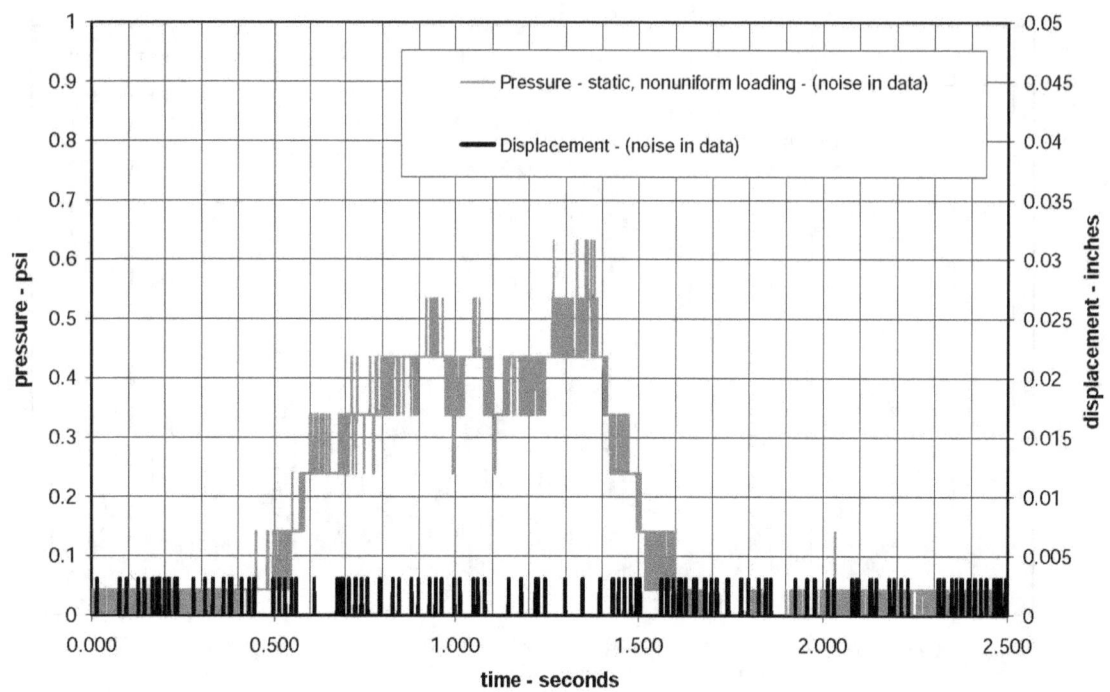

Figure A-53.—Category 3C - structure #1 - test 1 - static, nonuniform loading.
Hollow-core concrete-block ventilation stopping - LLEM test #427.

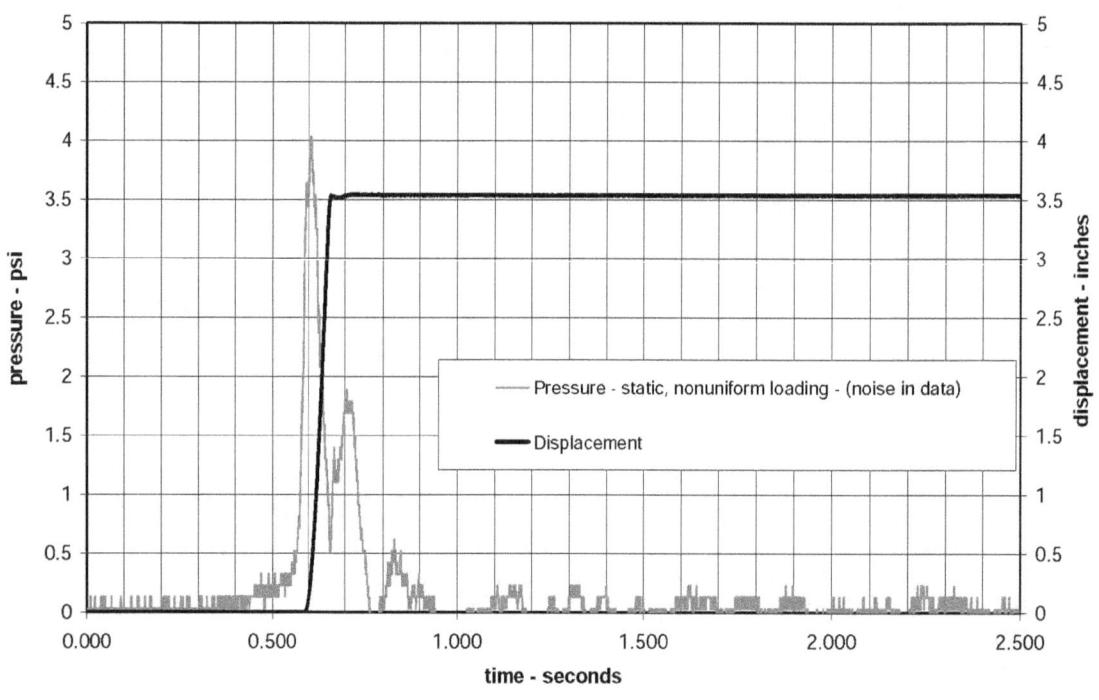

Figure A-54.—Category 3C - structure #1 - test 2 - static, nonuniform loading.
Hollow-core concrete-block ventilation stopping - LLEM test #428.

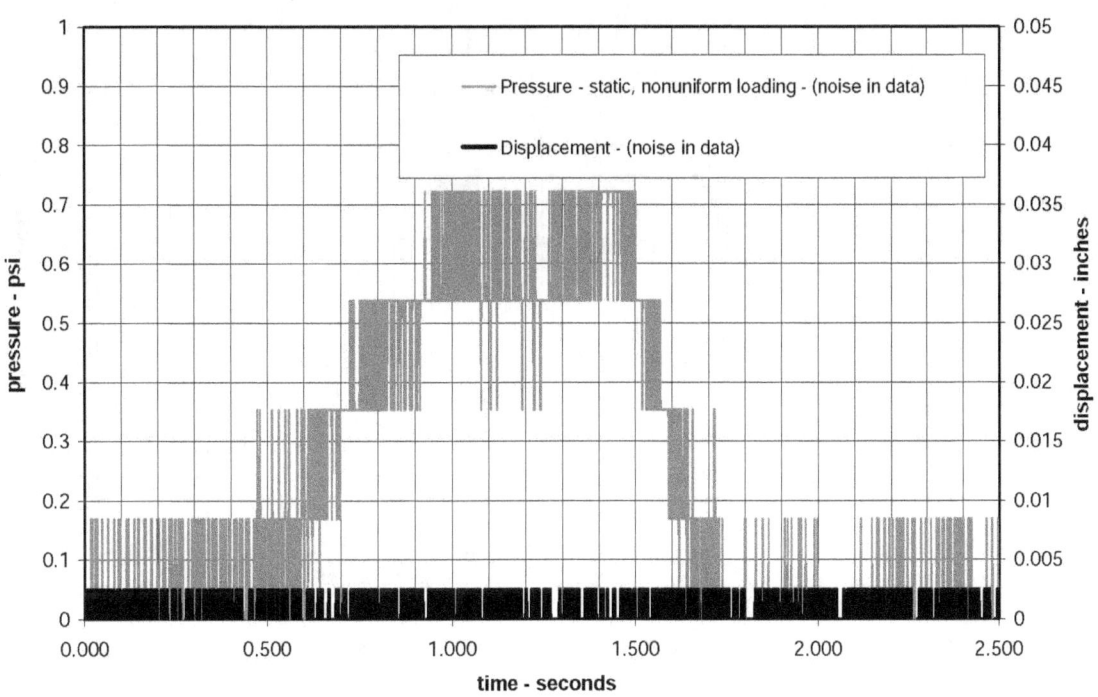

Figure A-55.—Category 3C - structure #2 - test 1 - static, nonuniform loading. Hollow-core concrete-block ventilation stopping - LLEM test #427.

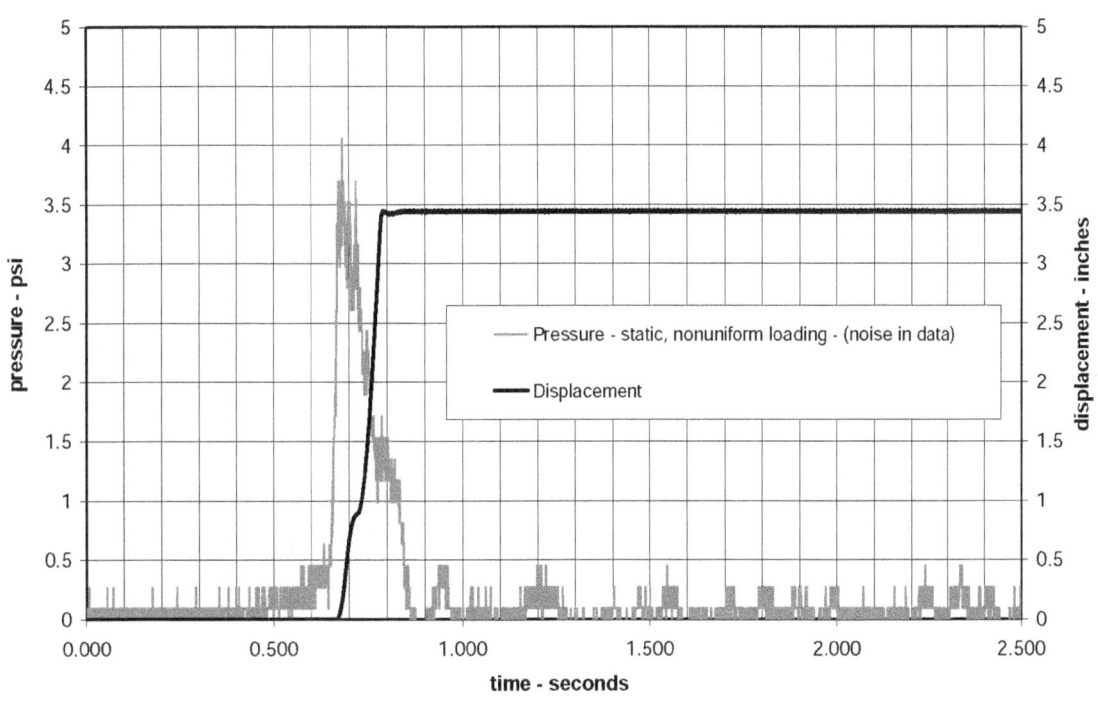

Figure A-56.—Category 3C - structure #2 - test 2 - static, nonuniform loading. Hollow-core concrete-block ventilation stopping - LLEM test #428.

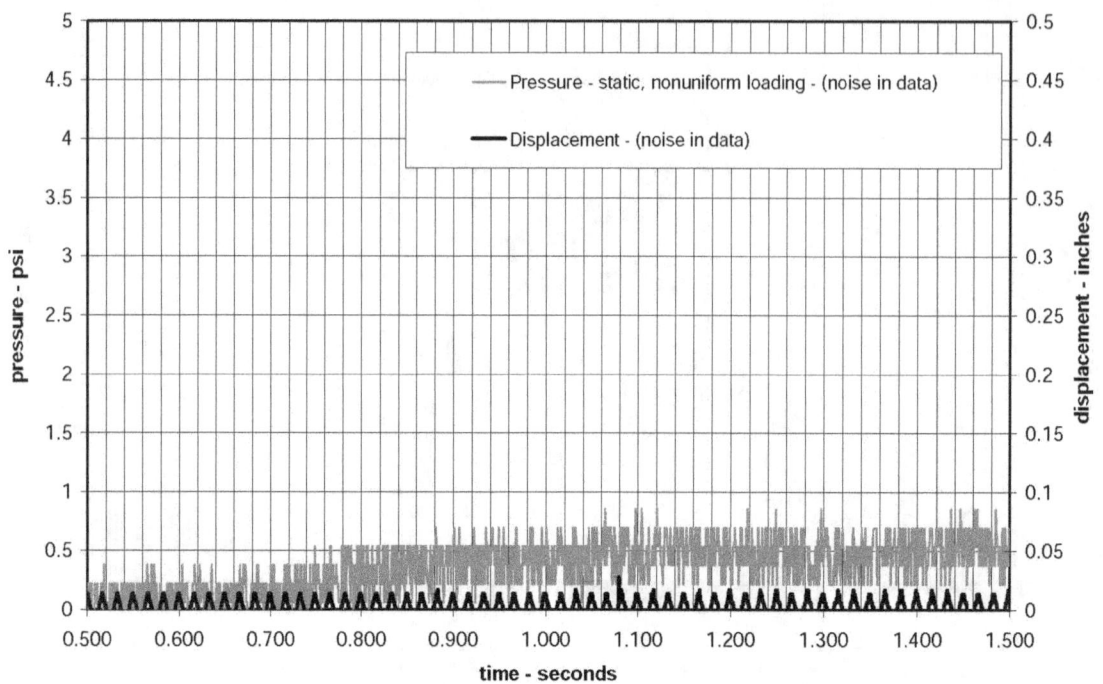

Figure A-57.—Category 3C - structure #3 - test 1 - static, nonuniform loading.
Hollow-core concrete-block ventilation stopping - LLEM test #427.

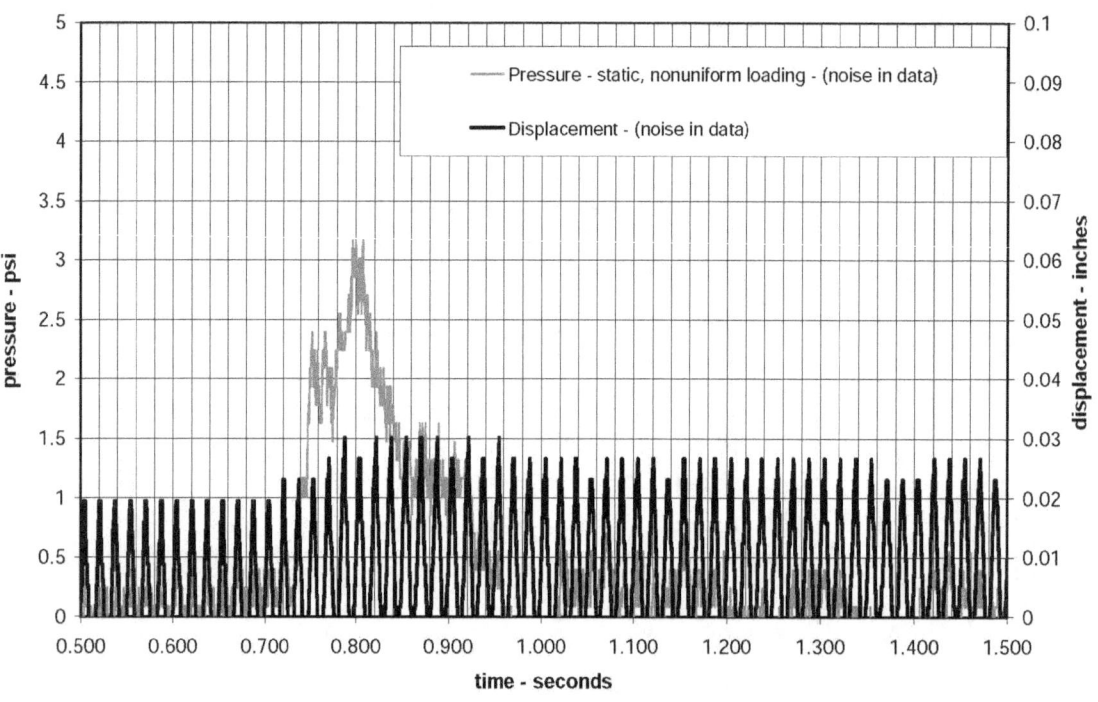

Figure A-58.—Category 3C - structure #3 - test 2 - static, nonuniform loading.
Hollow-core concrete-block ventilation stopping - LLEM test #428.

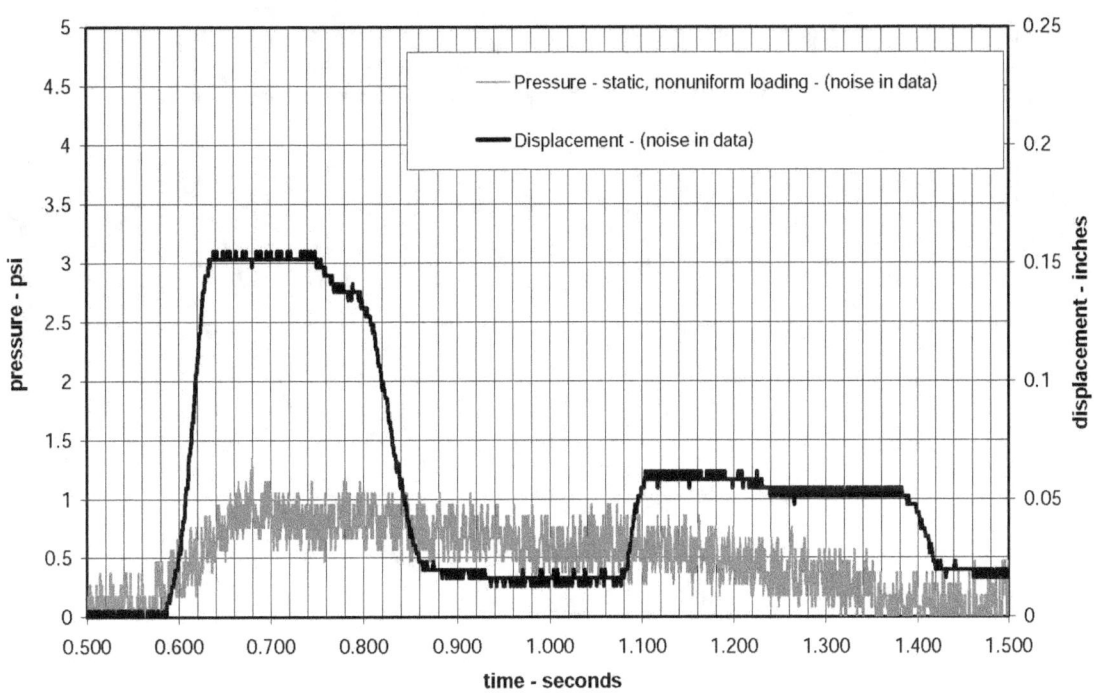

Figure A-59.—Category 3C - structure #3 - test 3 - static, nonuniform loading. Hollow-core concrete-block ventilation stopping - LLEM test #429.

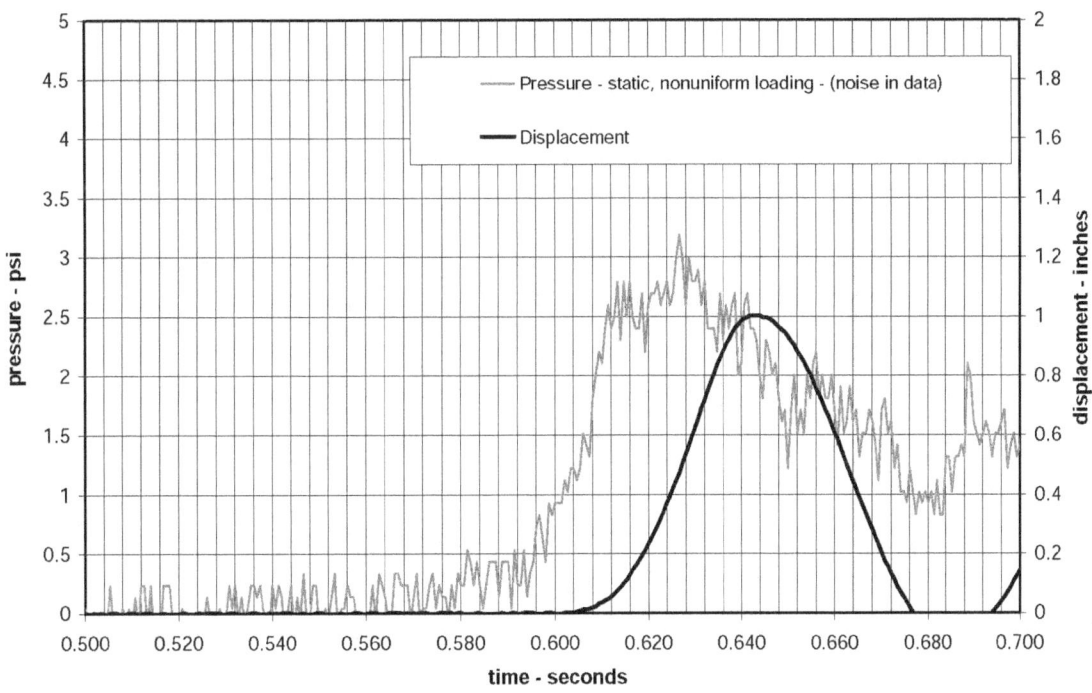

Figure A-60.—Category 3C - structure #3 - test 4 - static, nonuniform loading. Hollow-core concrete-block ventilation stopping - LLEM test #430.

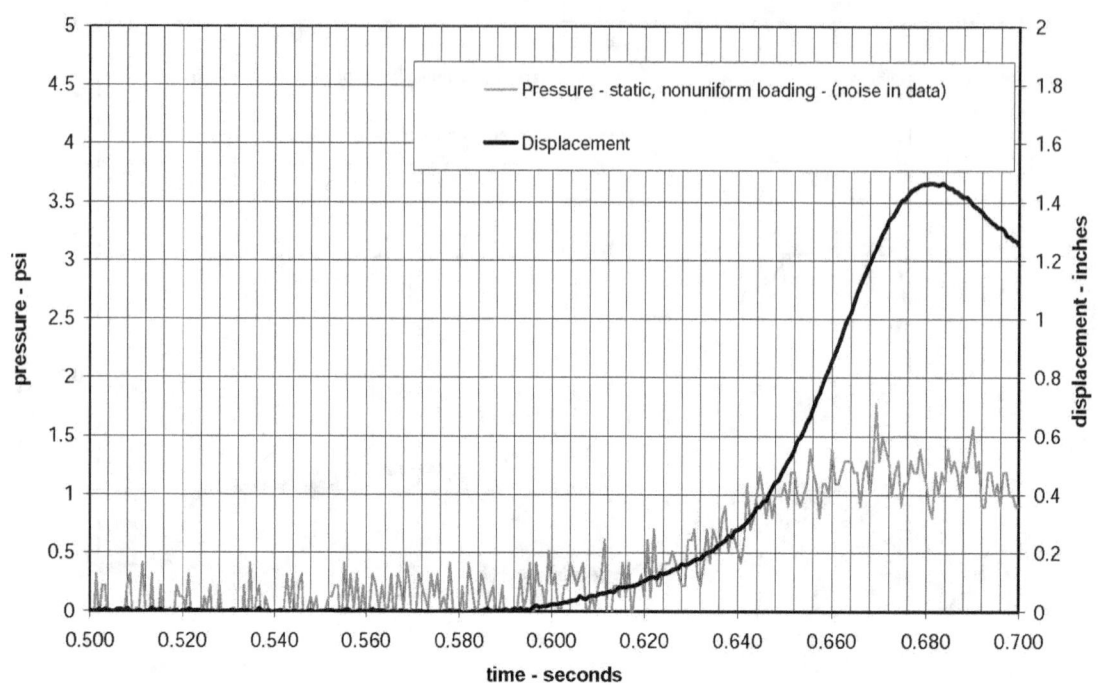

Figure A-61.—Category 3C - structure #3 - test 5 - static, nonuniform loading.
Hollow-core concrete-block ventilation stopping - LLEM test #432.

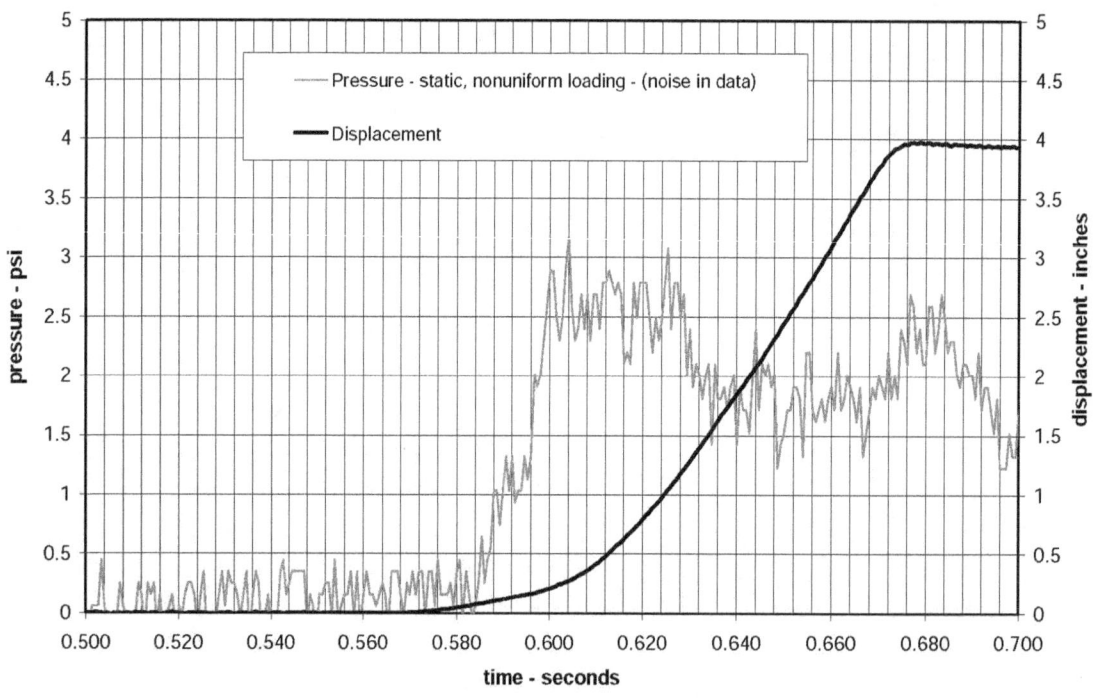

Figure A-62.—Category 3C - structure #3 - test 6 - static, nonuniform loading.
Hollow-core concrete-block ventilation stopping - LLEM test #433.

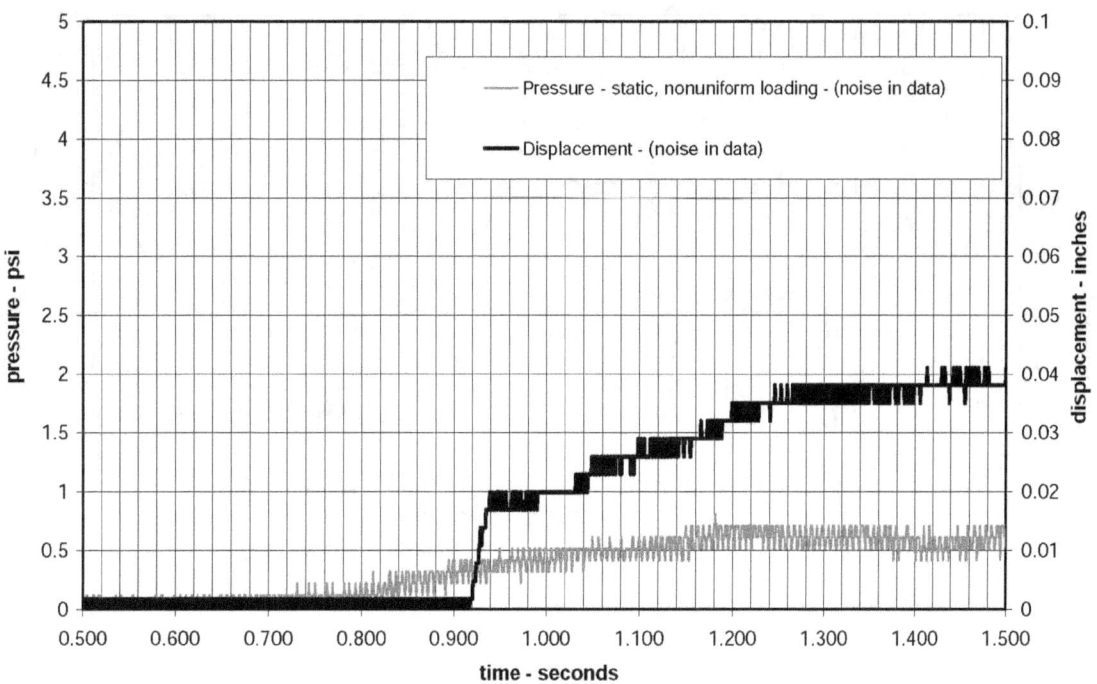

Figure A-63.—Category 3C - structure #4 - test 1 - static, nonuniform load. Hollow-core concrete-block ventilation stopping - LLEM test #427.

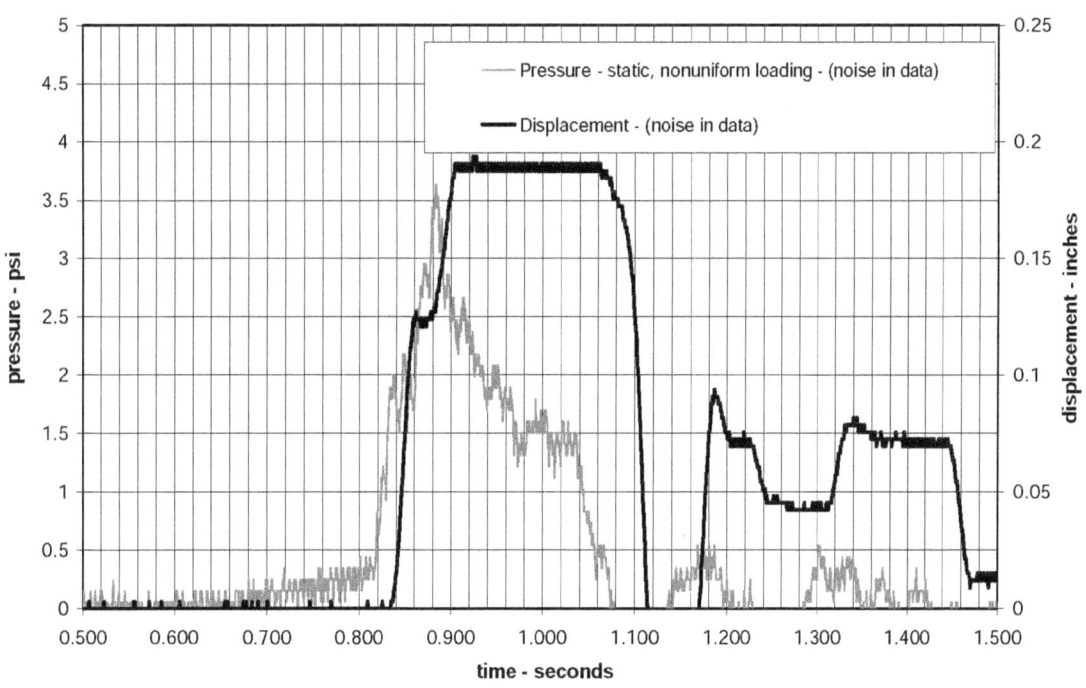

Figure A-64.—Category 3C - structure #4 - test 2 - static, nonuniform load. Hollow-core concrete-block ventilation stopping - LLEM test #428.

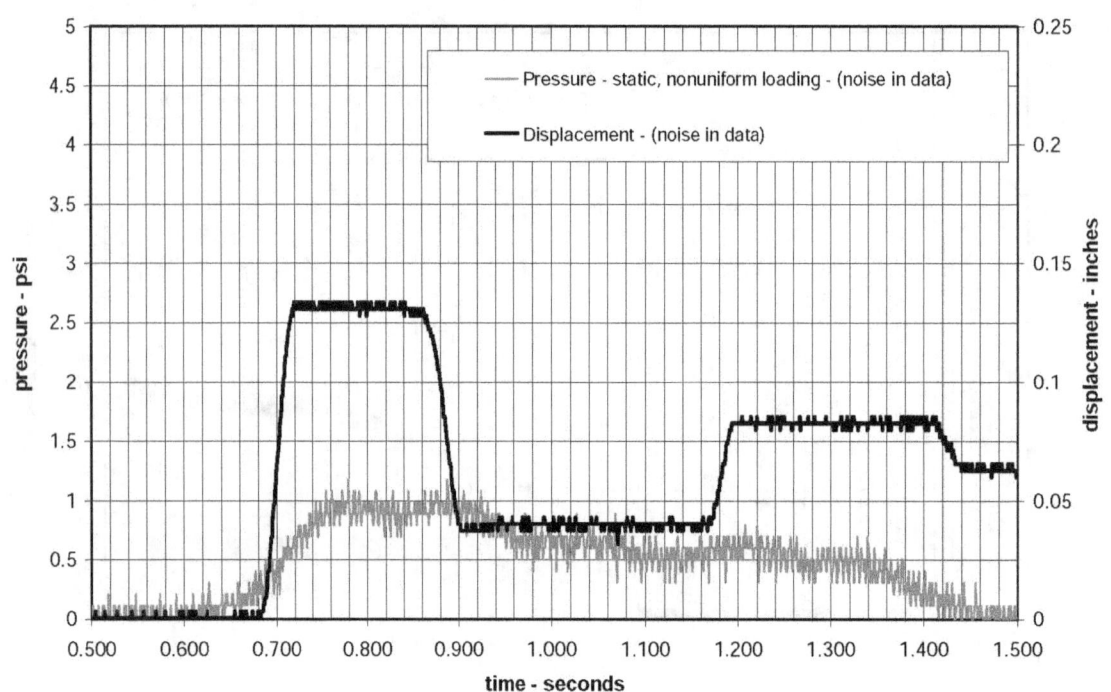

Figure A-65.—Category 3C - structure #4 - test 3 - static, nonuniform load. Hollow-core concrete-block ventilation stopping - LLEM test #429.

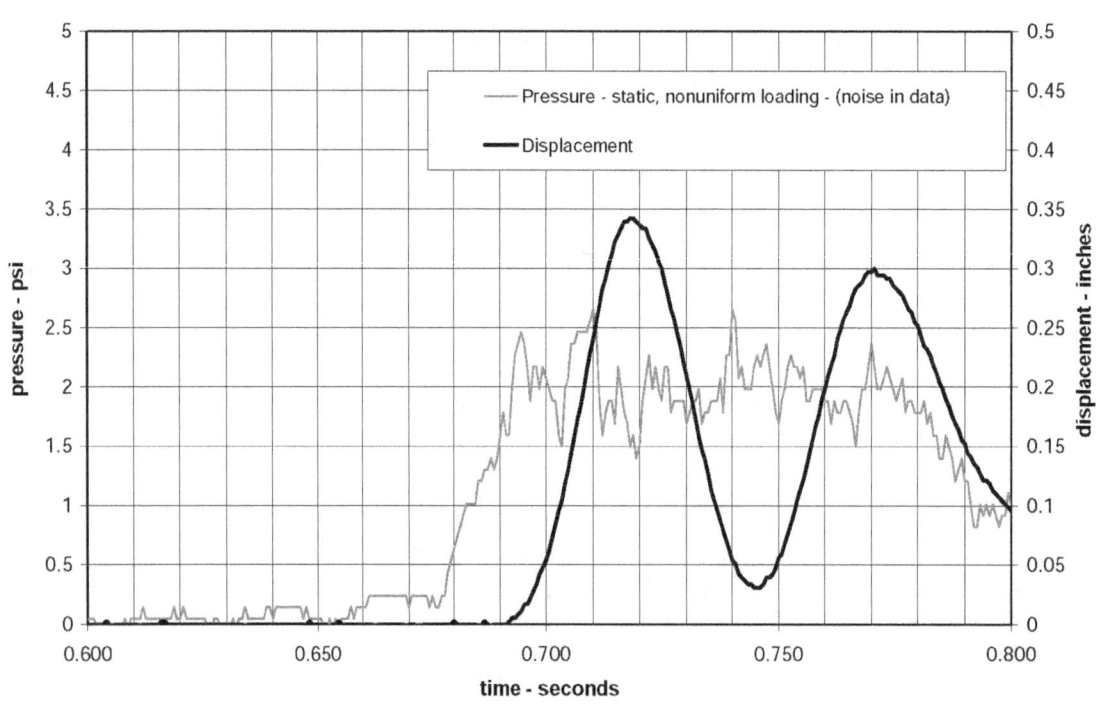

Figure A-66.—Category 3C - structure #4 - test 4 - static, nonuniform load. Hollow-core concrete-block ventilation stopping - LLEM test #430.

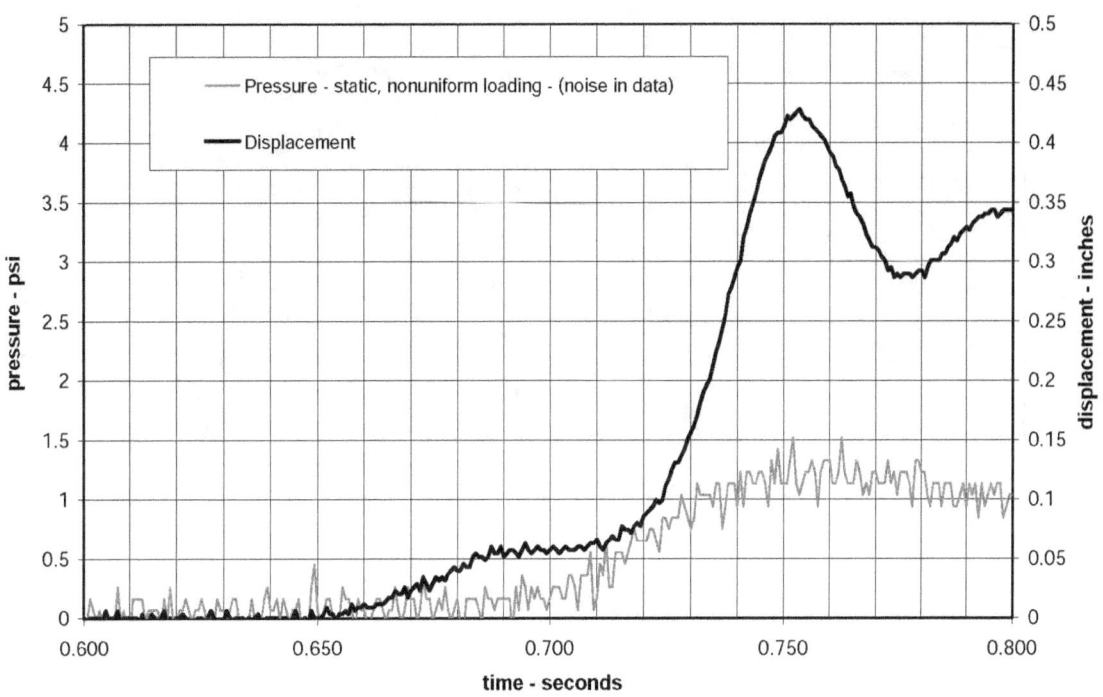

Figure A-67.—Category 3C - structure #4 - test 5 - static, nonuniform load. Hollow-core concrete-block ventilation stopping - LLEM test #432.

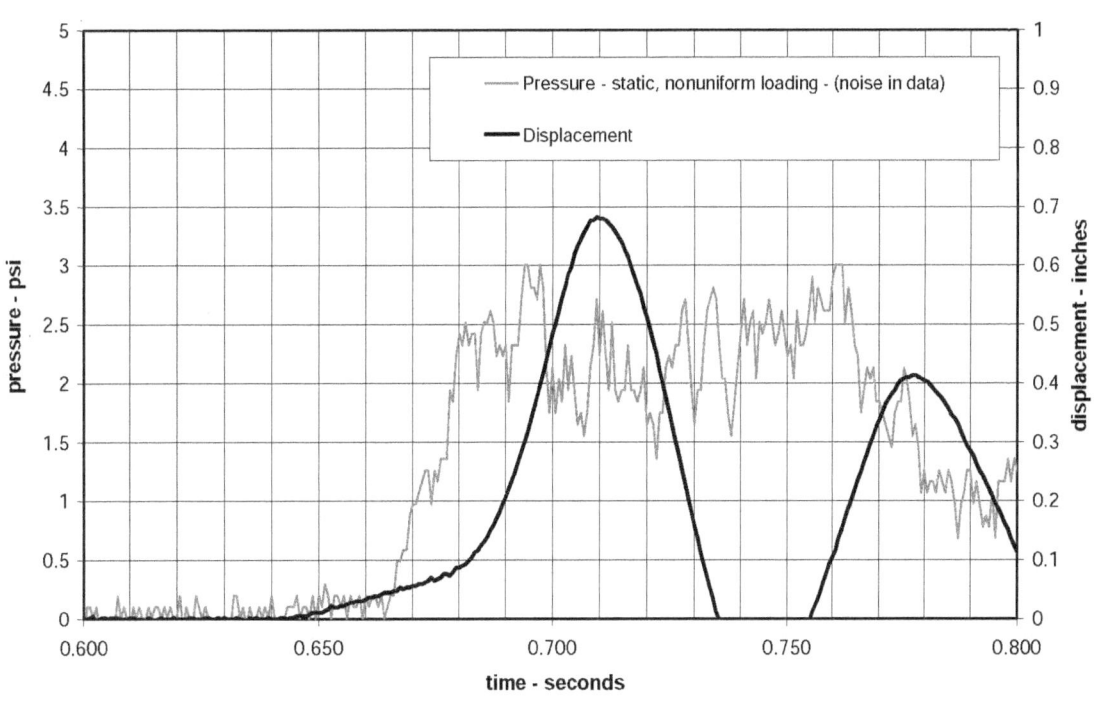

Figure A-68.—Category 3C - structure #4 - test 6 - static, nonuniform load. Hollow-core concrete-block ventilation stopping - LLEM test #433.

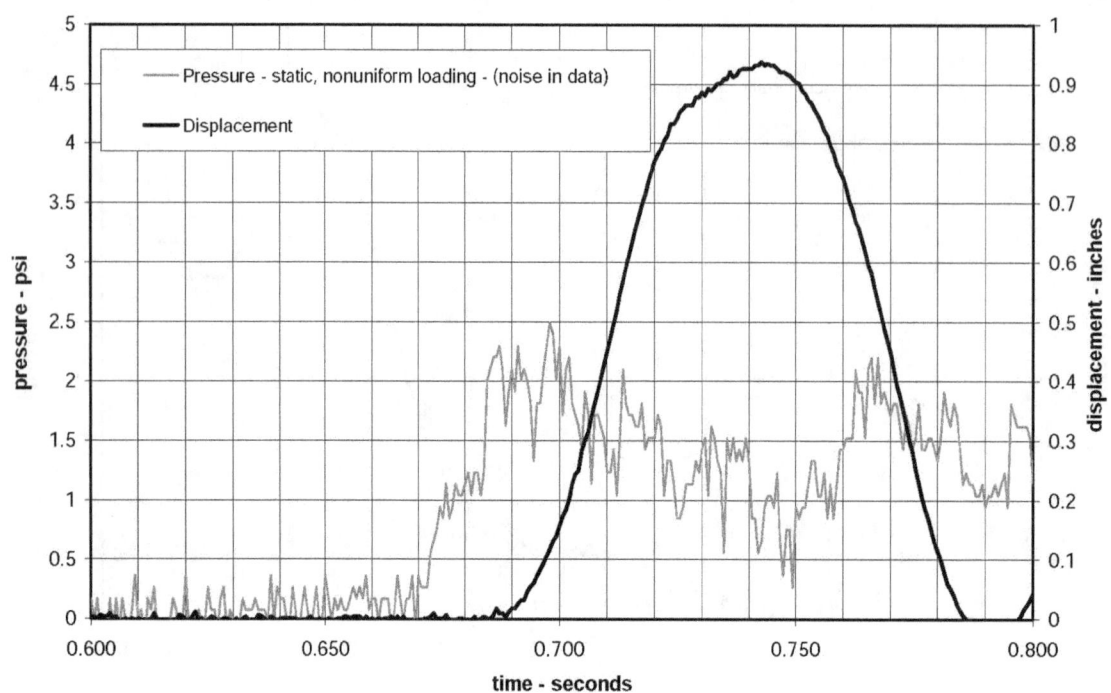

Figure A-69.—Category 3C - structure #4 - test 7 - static, nonuniform load.
Hollow-core concrete-block ventilation stopping - LLEM test #434.

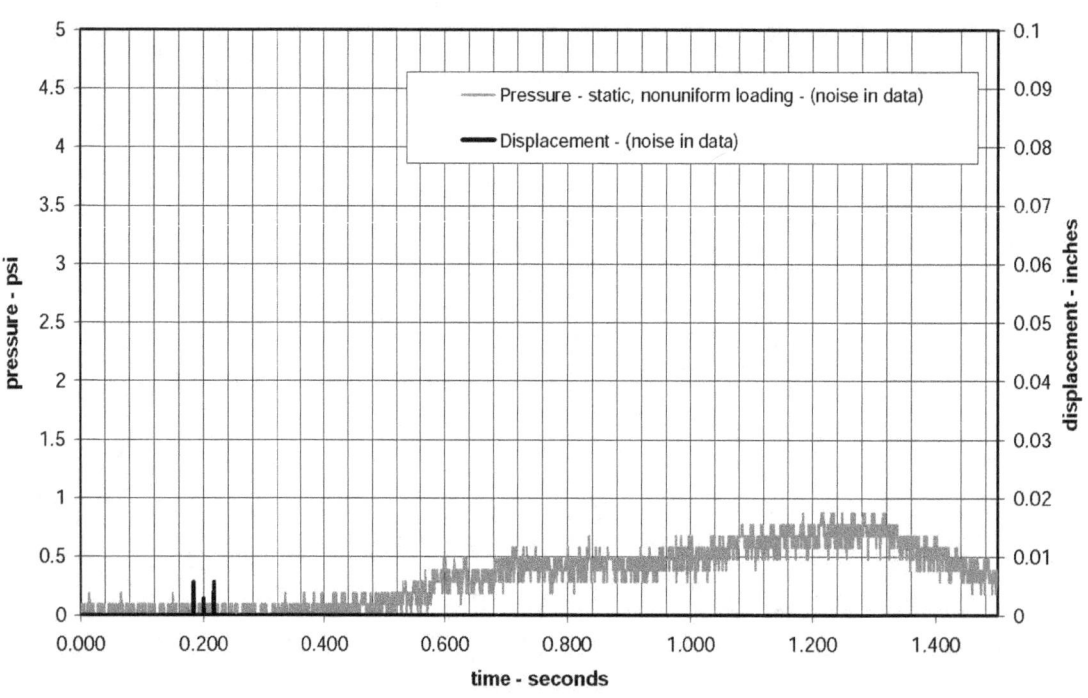

Figure A-70.—Category 3C - structure #5 - test 1 - static, nonuniform loading.
Solid-concrete-block ventilation stopping - LLEM test #457.

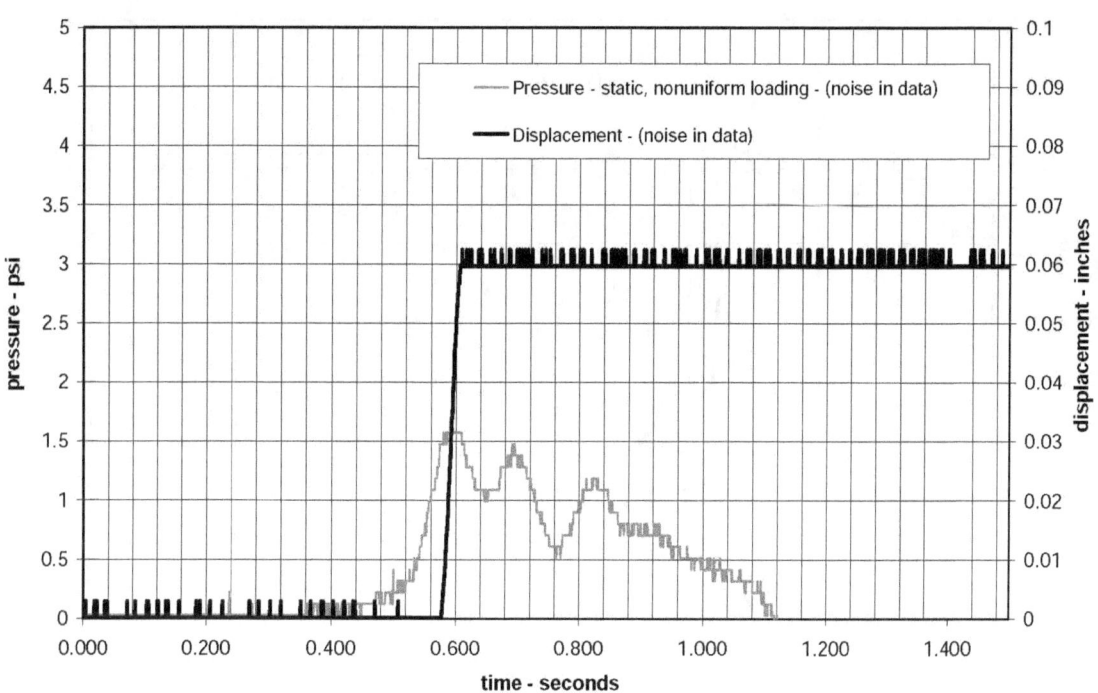

Figure A-71.—Category 3C - structure #5 - test 2 - static, nonuniform loading. Solid-concrete-block ventilation stopping - LLEM test #458.

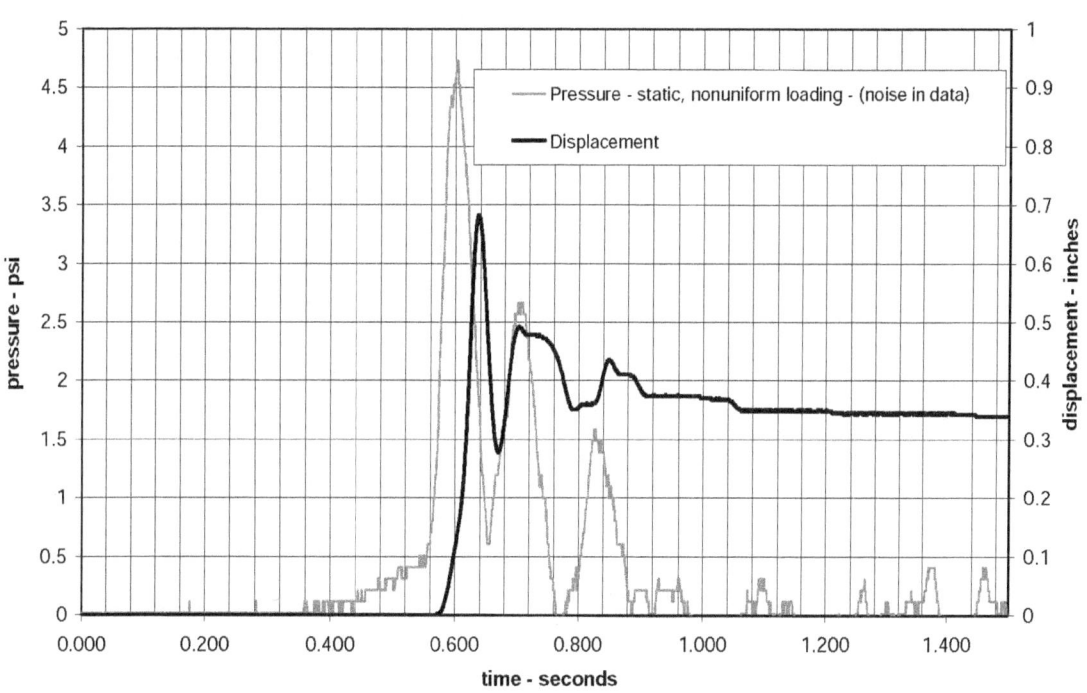

Figure A-72.—Category 3C - structure #5 - test 3 - static, nonuniform loading. Solid-concrete-block ventilation stopping - LLEM test #459.

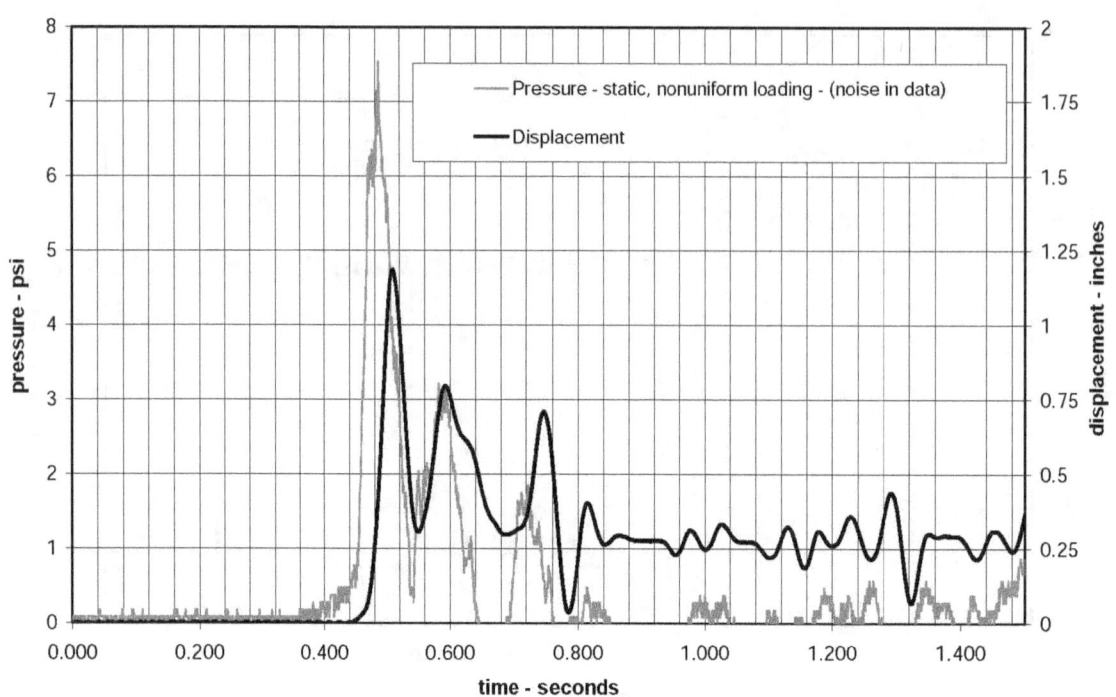

Figure A-73.—Category 3C - structure #5 - test 4 - static, nonuniform loading. Solid-concrete-block ventilation stopping - LLEM test #460.

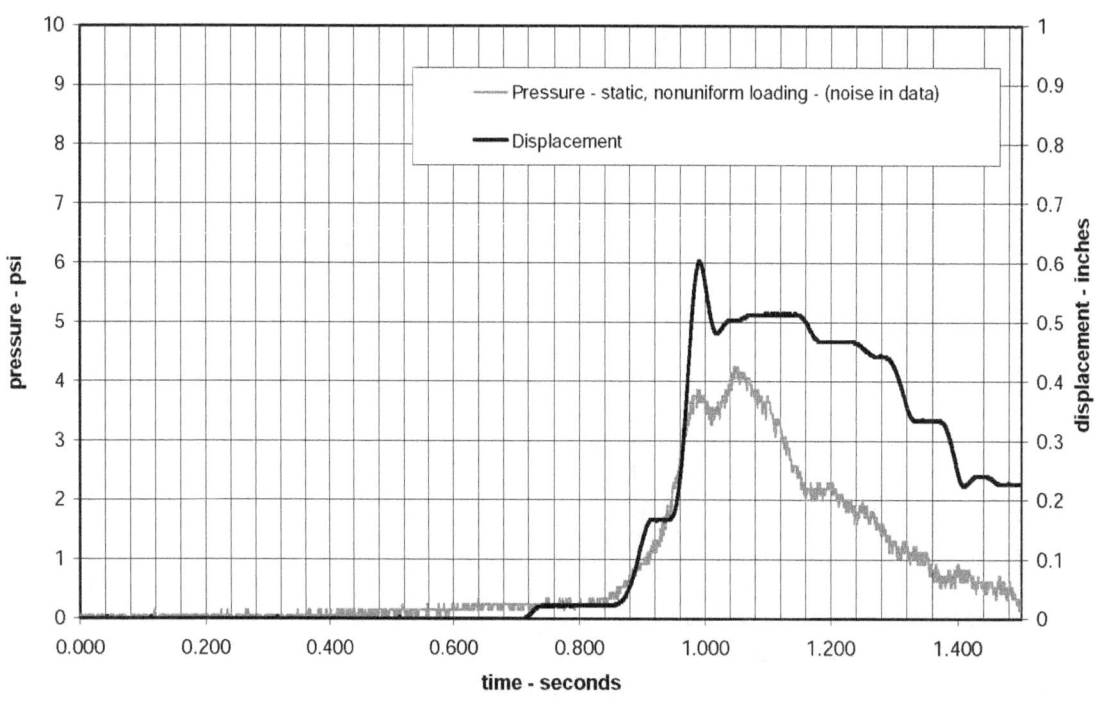

Figure A-74.—Category 3C - structure #5 - test 5 - static, nonuniform loading. Solid-concrete-block ventilation stopping - LLEM test #461.

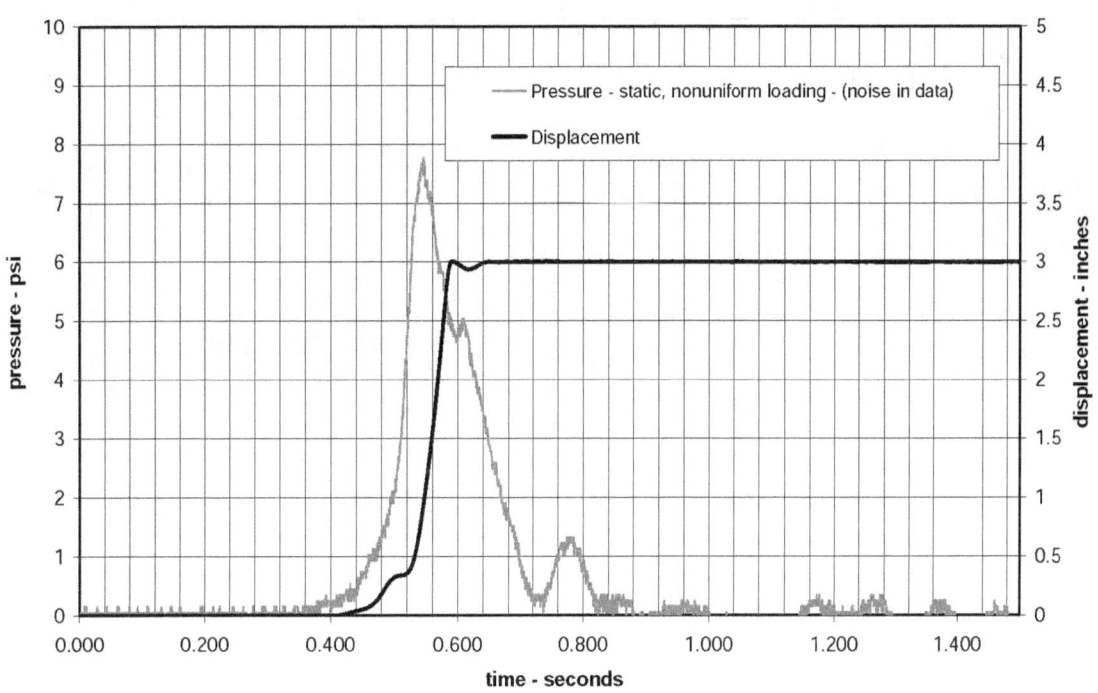

Figure A-75.—Category 3C - structure #5 - test 6 - static, nonuniform loading.
Solid-concrete-block ventilation stopping - LLEM test #462.

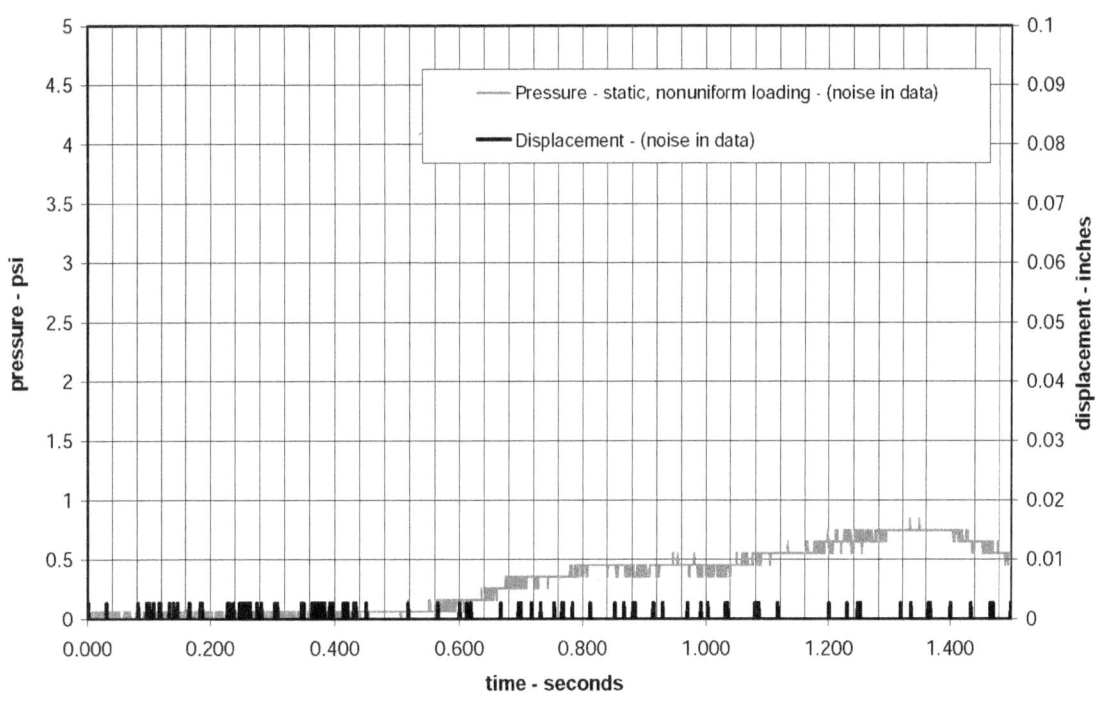

Figure A-76.—Category 3C - structure #6 - test 1 - static, nonuniform loading.
Solid-concrete-block ventilation stopping - LLEM test #457.

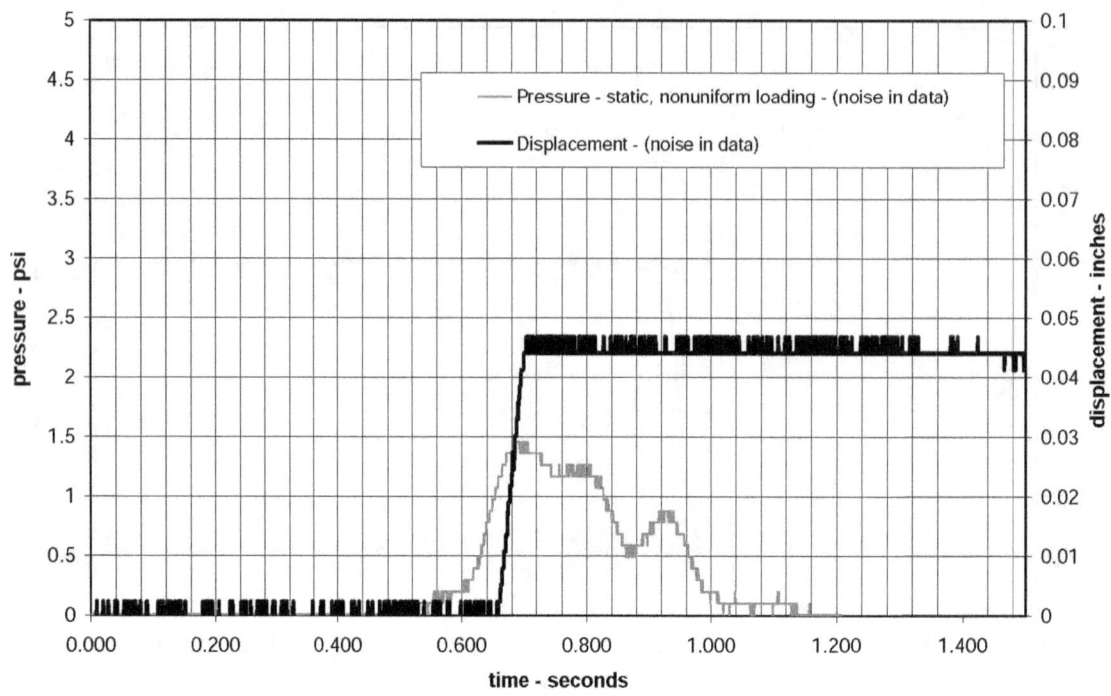

Figure A-77.—Category 3C - structure #6 - test 2 - static, nonuniform loading.
Solid-concrete-block ventilation stopping - LLEM test #458.

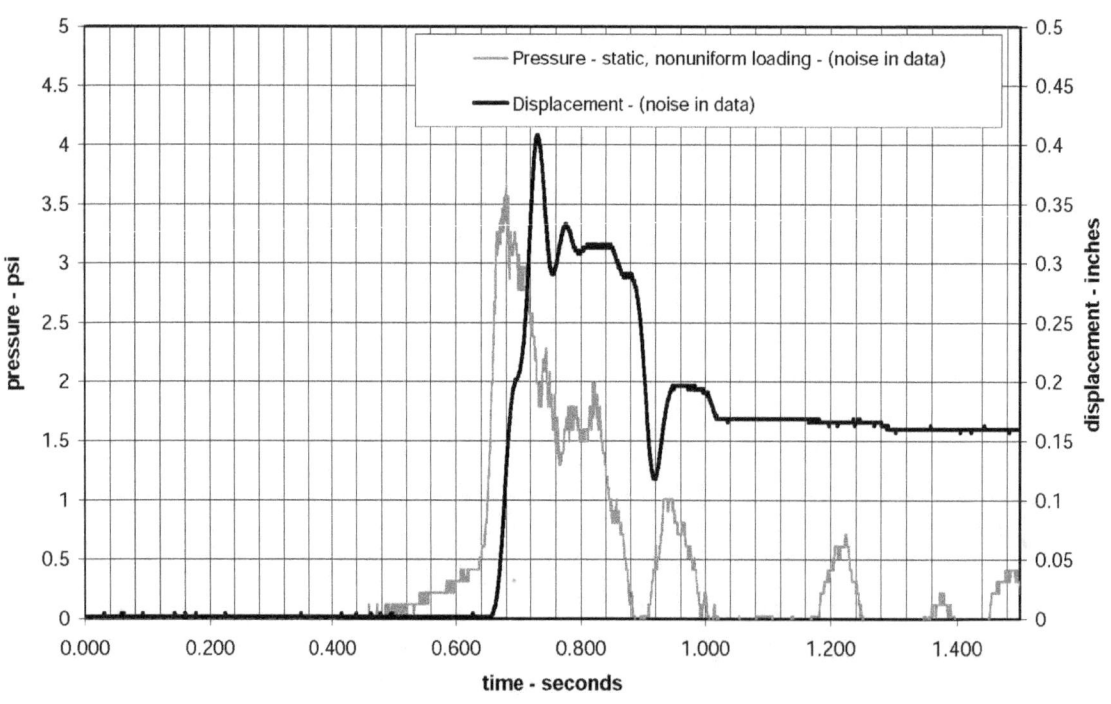

Figure A-78.—Category 3C - structure #6 - test 3 - static, nonuniform loading.
Solid-concrete-block ventilation stopping - LLEM test #459.

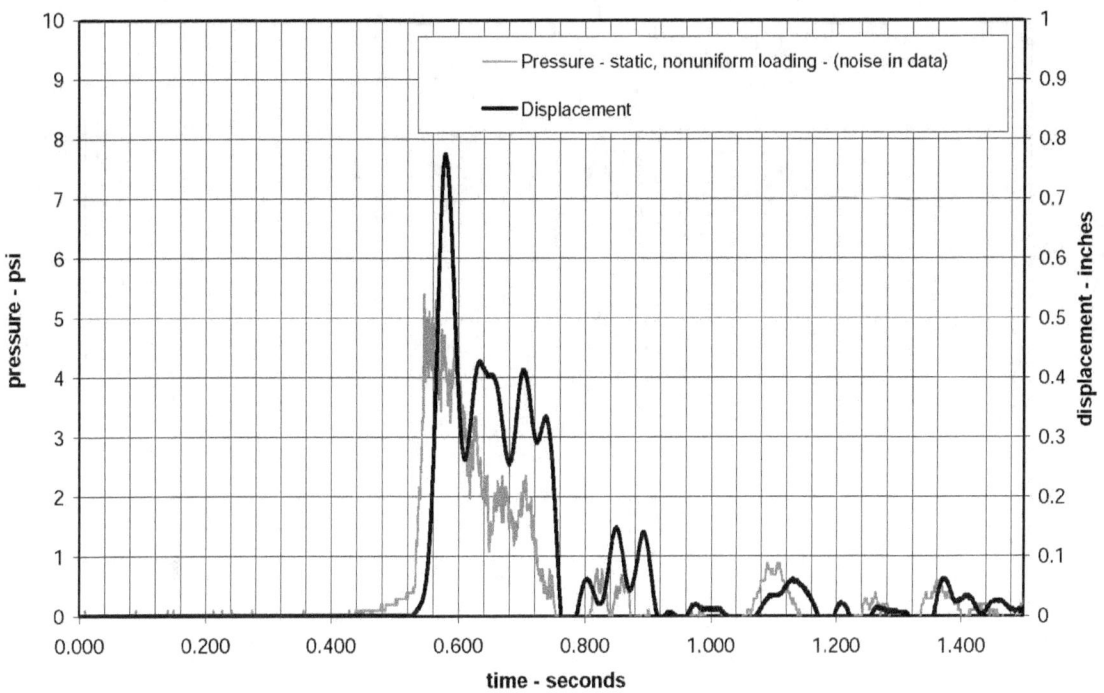

Figure A-79.—Category 3C - structure #6 - test 4 - static, nonuniform loading. Solid-concrete-block ventilation stopping - LLEM test #460.

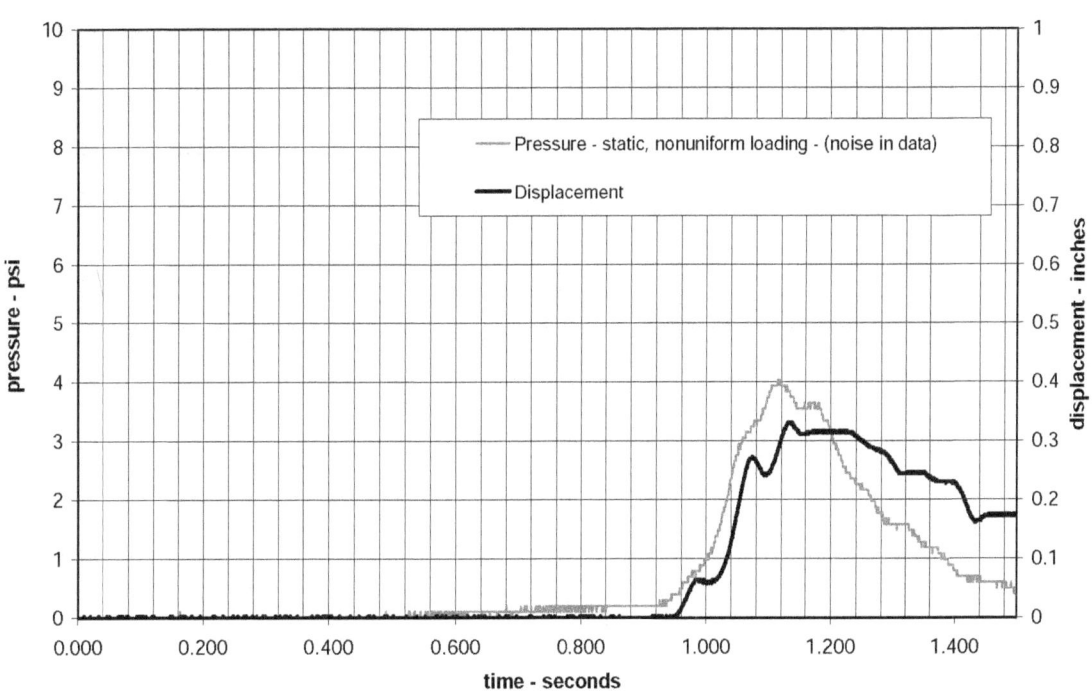

Figure A-80.—Category 3C - structure #6 - test 5 - static, nonuniform loading. Solid-concrete-block ventilation stopping - LLEM test #461.

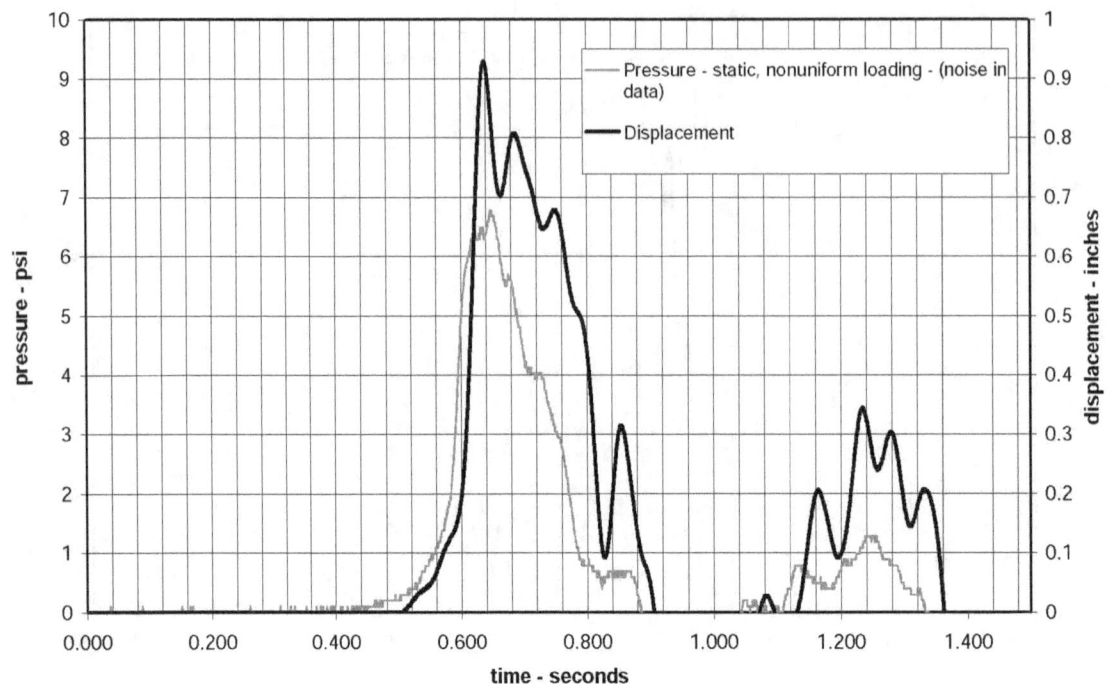

Figure A-81.—Category 3C - structure #6 - test 6 - static, nonuniform loading. Solid-concrete-block ventilation stopping - LLEM test #462.

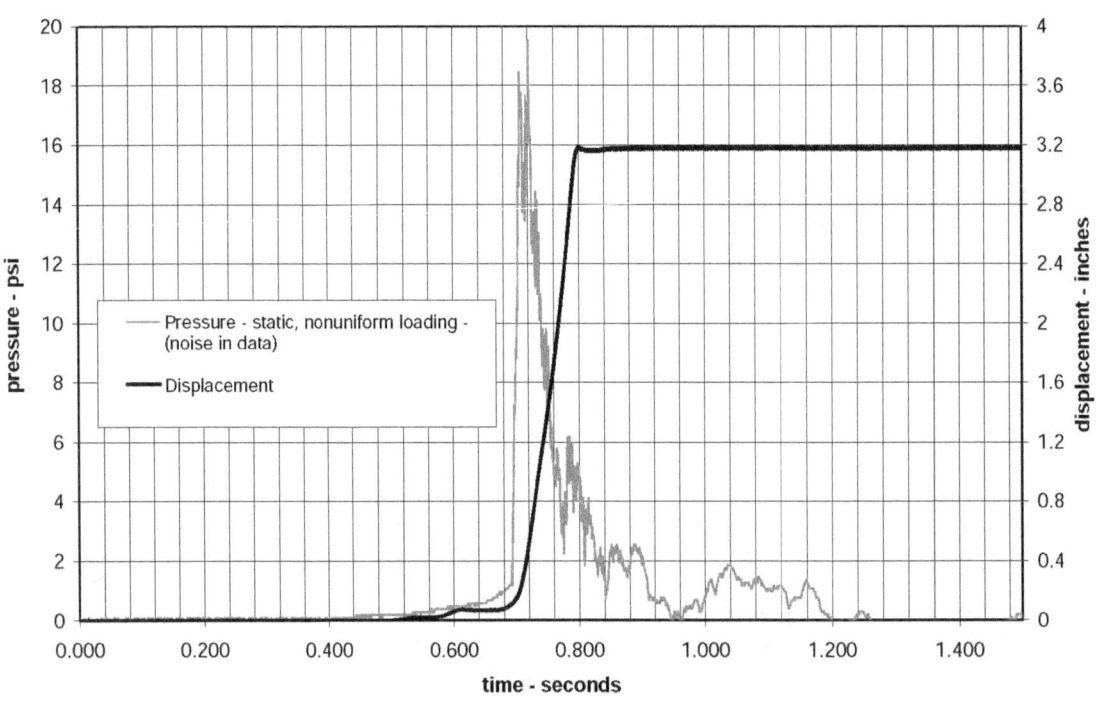

Figure A-82.—Category 3C - structure #6 - test 7 - static, nonuniform loading. Solid-concrete-block ventilation stopping - LLEM test #463.

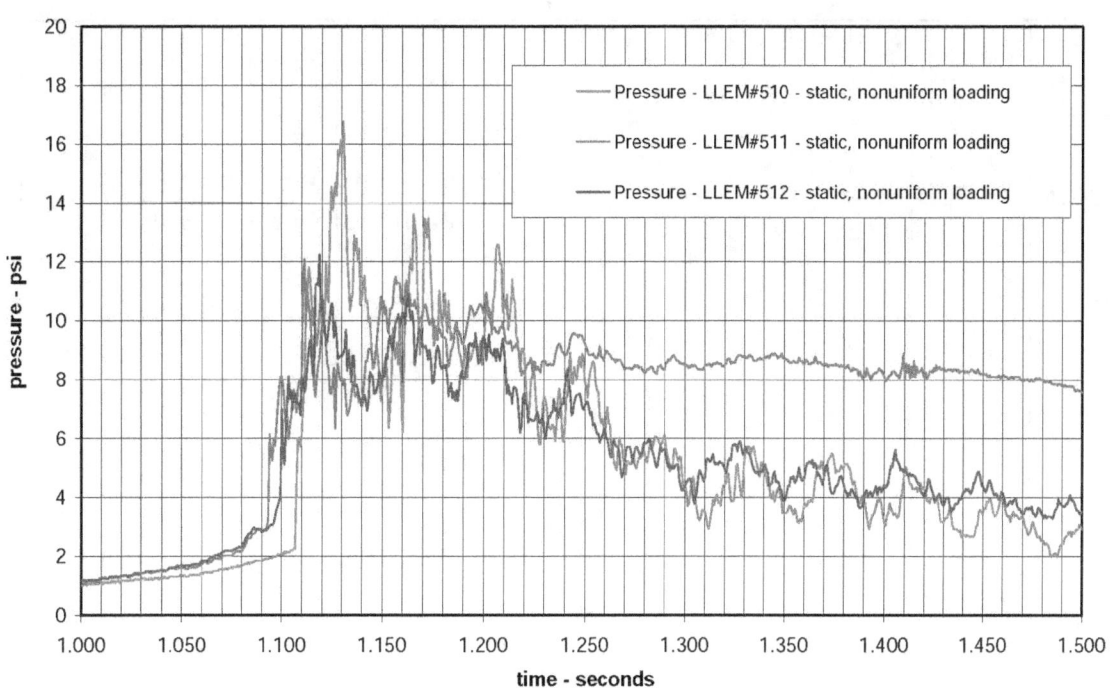

Figure A-83.—Category 3C - structure #7 - tests 1 to 3 - static, nonuniform loading. Solid-concrete-block ventilation stopping - LLEM tests #510–512.

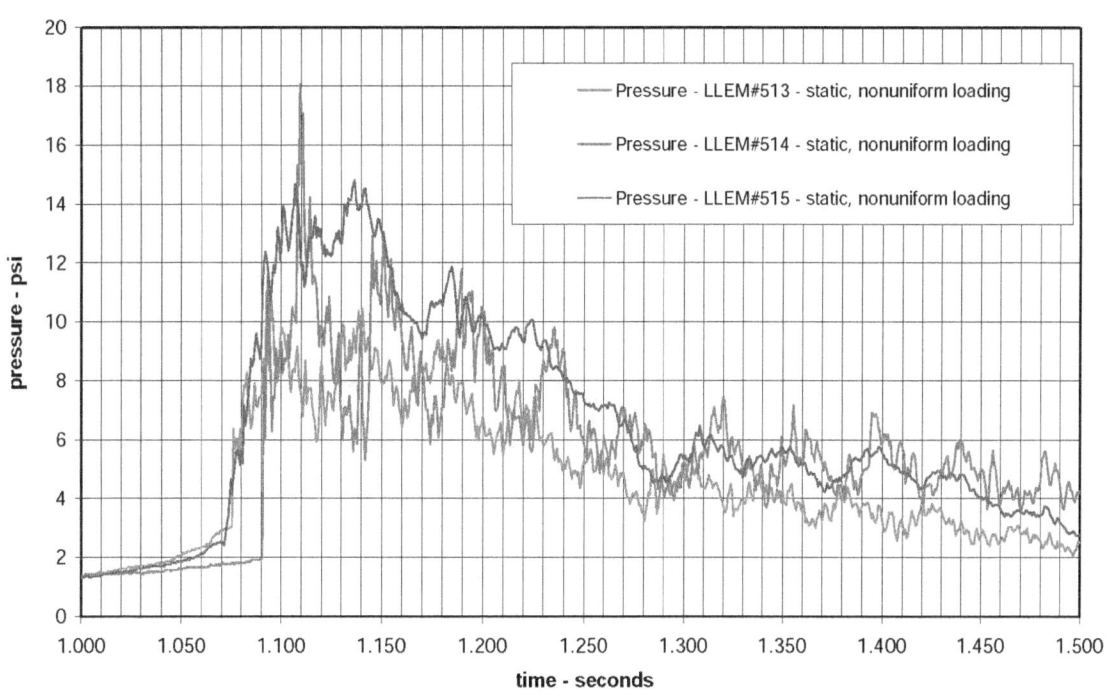

Figure A-84.—Category 3C - structure #7 - tests 4 to 6 - static, nonuniform loading. Solid-concrete-block ventilation stopping - LLEM tests #513–515.

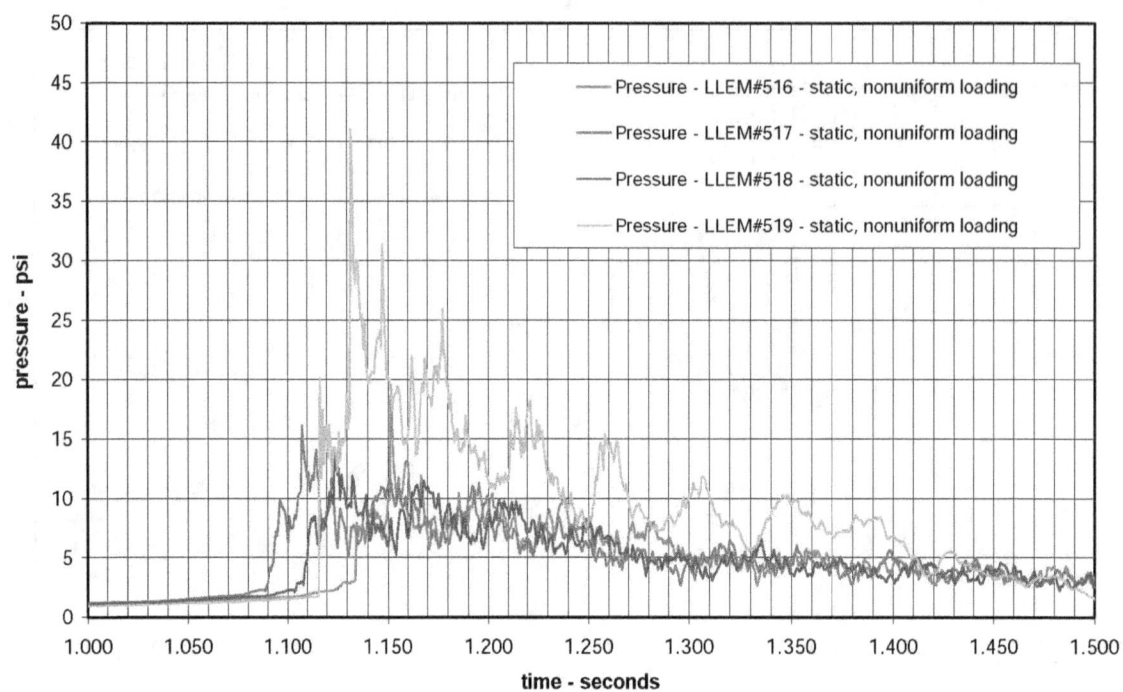

Figure A-85.—Category 3C - structure #7 - tests 7 to 10 - static, nonuniform loading. Solid-concrete-block ventilation stopping - LLEM tests #516–519.

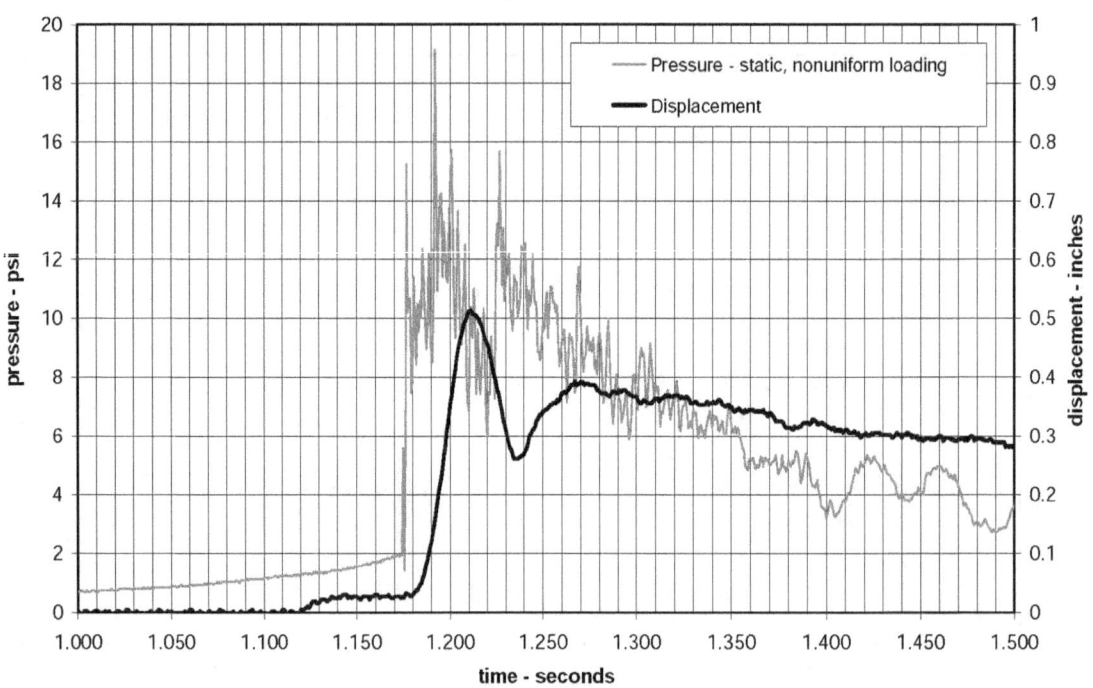

Figure A-86.—Category 3C - structure #8 - test 1 - static, nonuniform loading. Solid-concrete-block ventilation stopping - LLEM test #510.

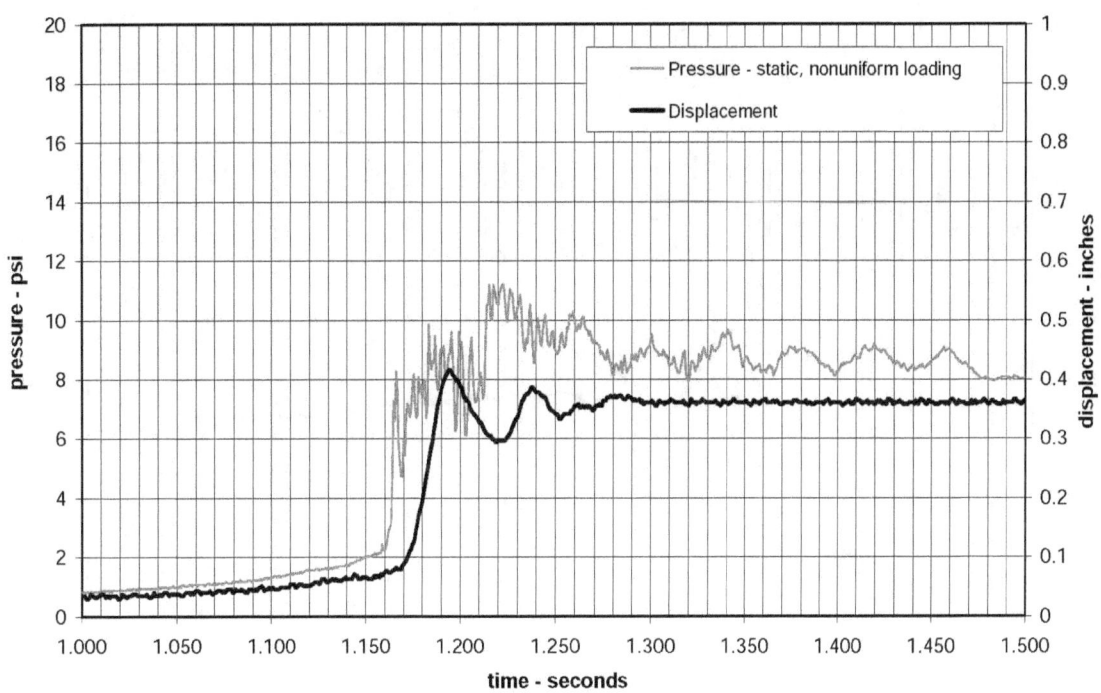

Figure A-87.—Category 3C - structure #8 - test 2 - static, nonuniform loading. Solid-concrete-block ventilation stopping - LLEM test #511.

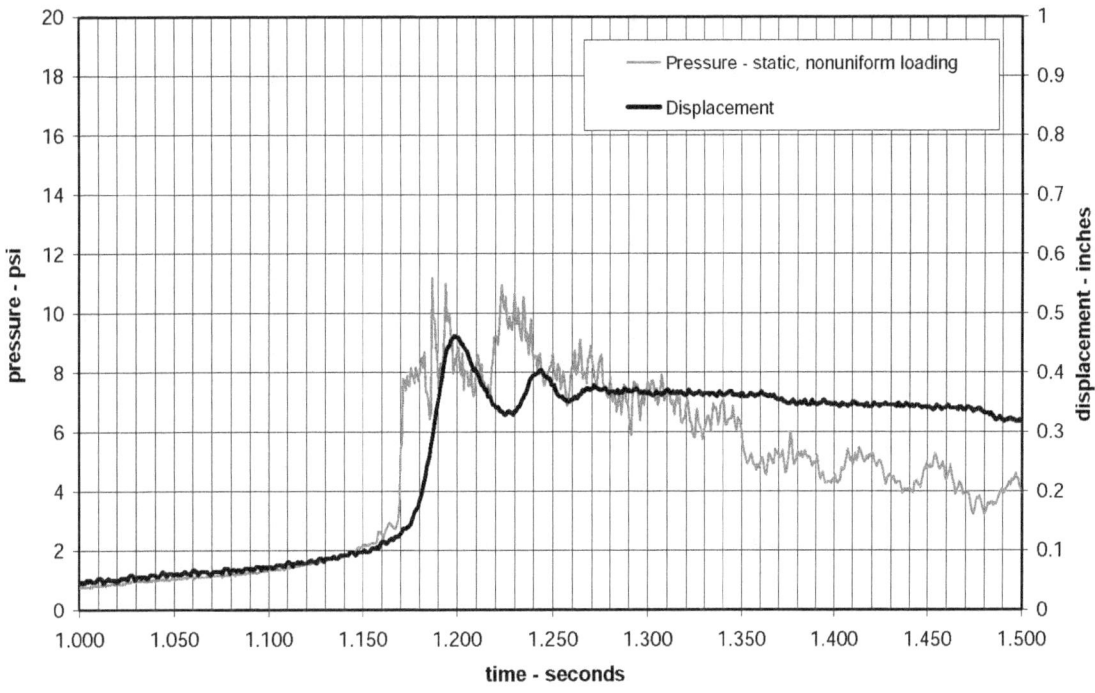

Figure A-88.—Category 3C - structure #8 - test 3 - static, nonuniform loading. Solid-concrete-block ventilation stopping - LLEM test #512.

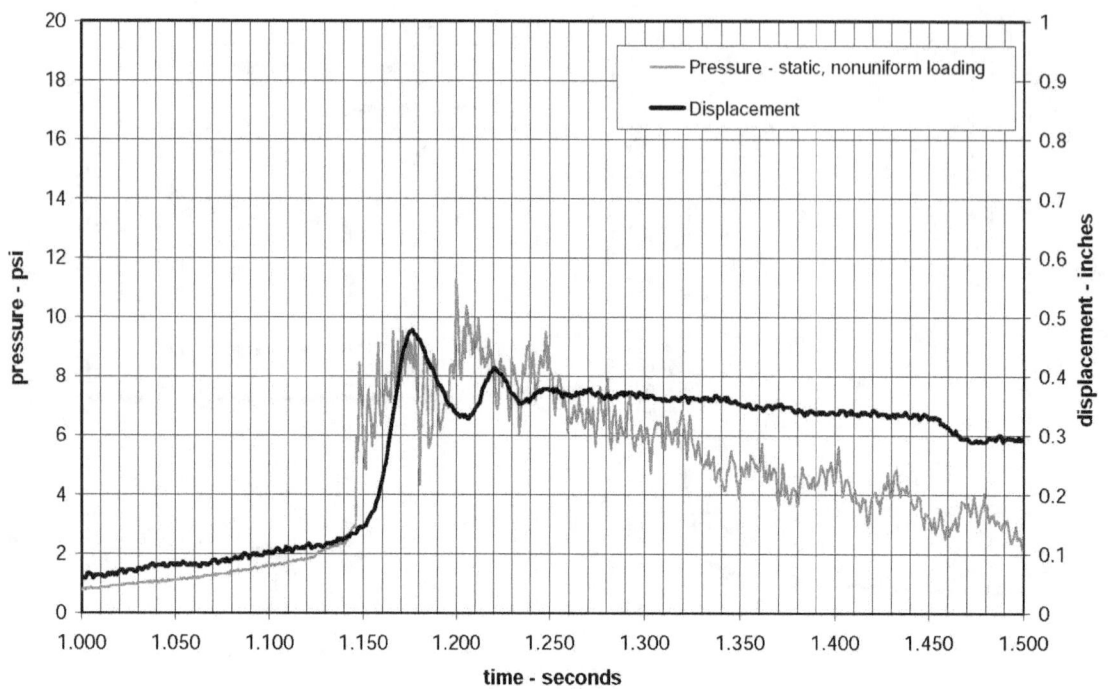

Figure A-89.—Category 3C - structure #8 - test 4 - static, nonuniform loading.
Solid-concrete-block ventilation stopping - LLEM test #513.

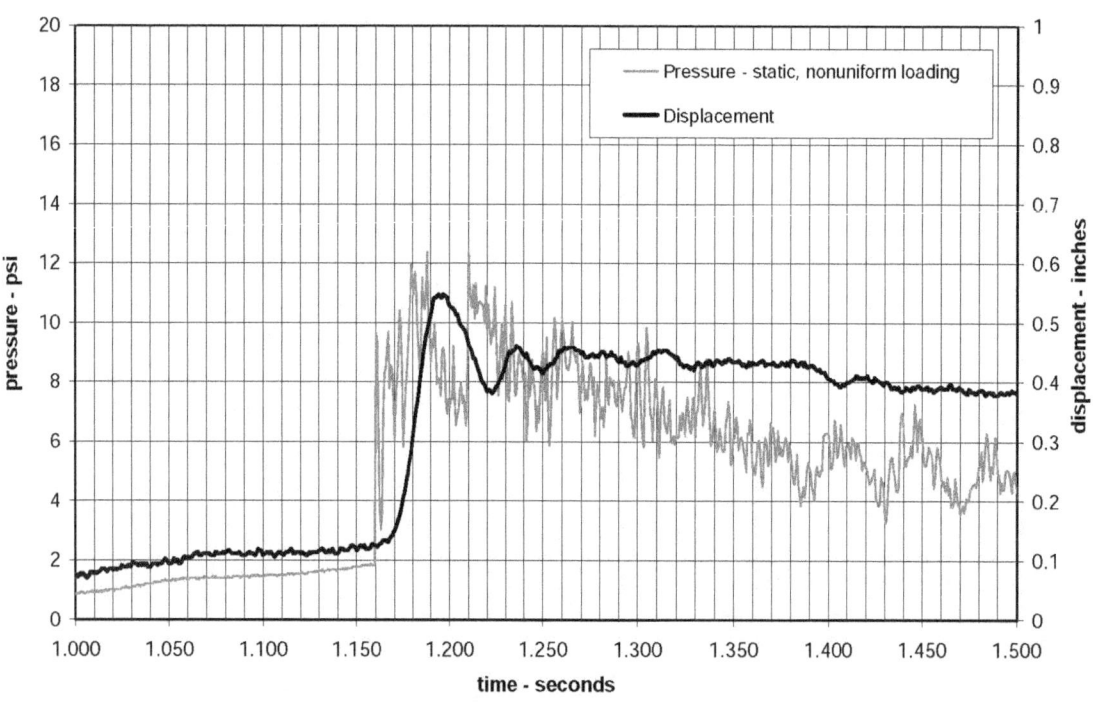

Figure A-90.—Category 3C - structure #8 - test 5 - static, nonuniform loading.
Solid-concrete-block ventilation stopping - LLEM test #514.

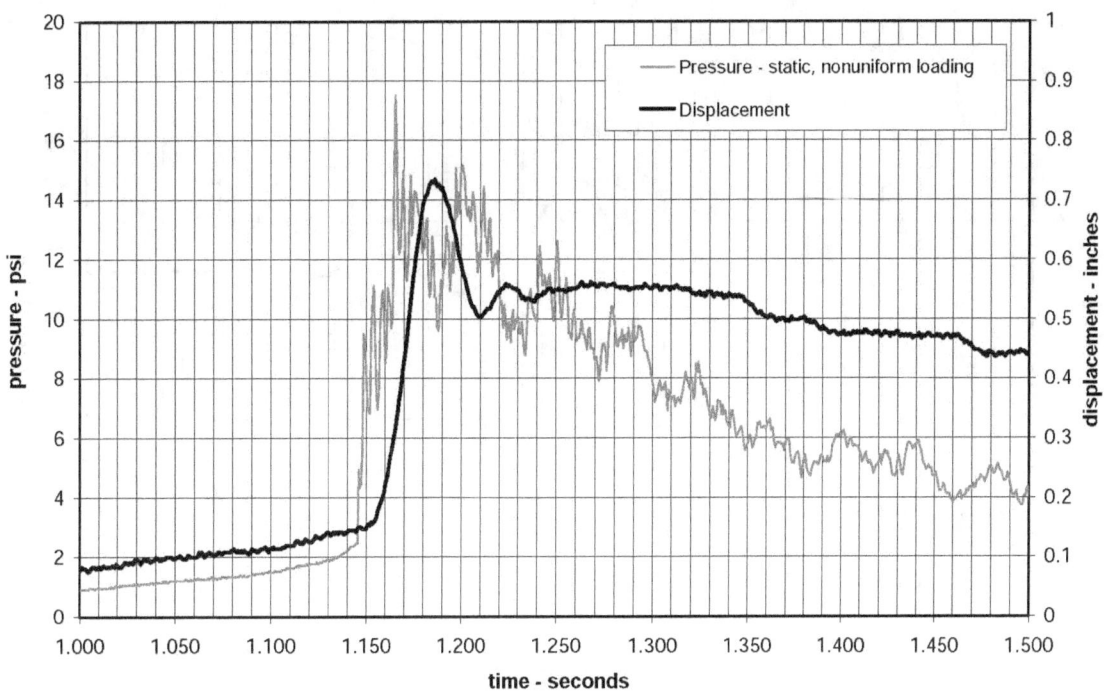

Figure A-91.—Category 3C - structure #8 - test 6 - static, nonuniform loading. Solid-concrete-block ventilation stopping - LLEM test #515.

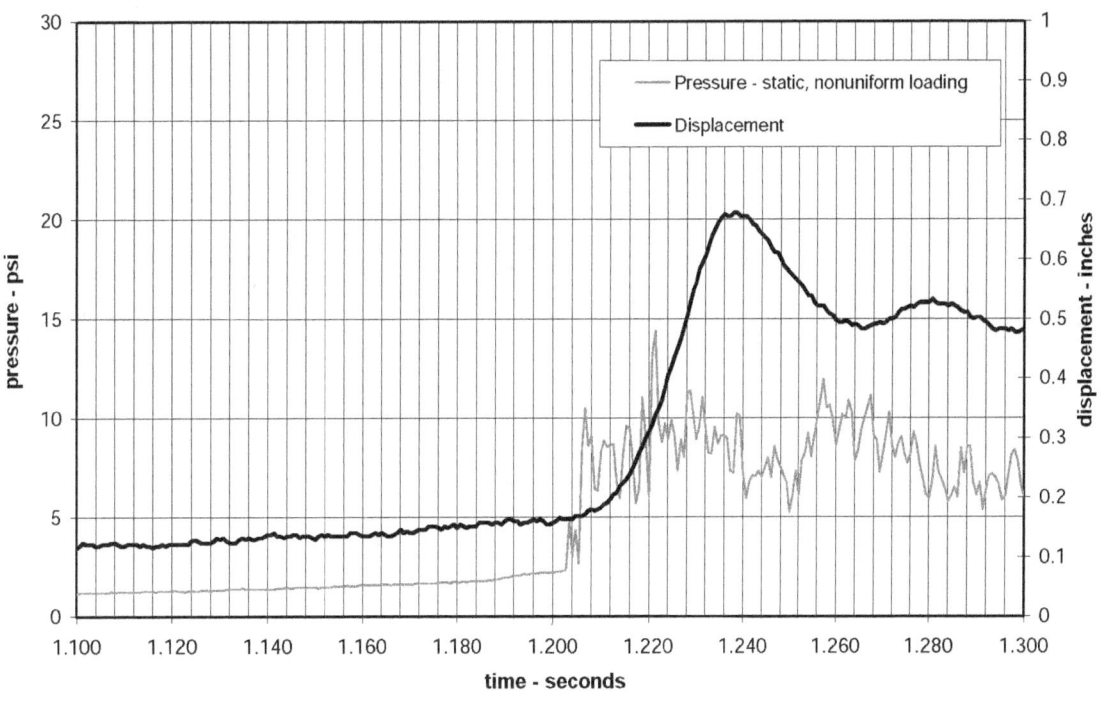

Figure A-92.—Category 3C - structure #8 - test 7 - static, nonuniform loading. Solid-concrete-block ventilation stopping - LLEM test #516.

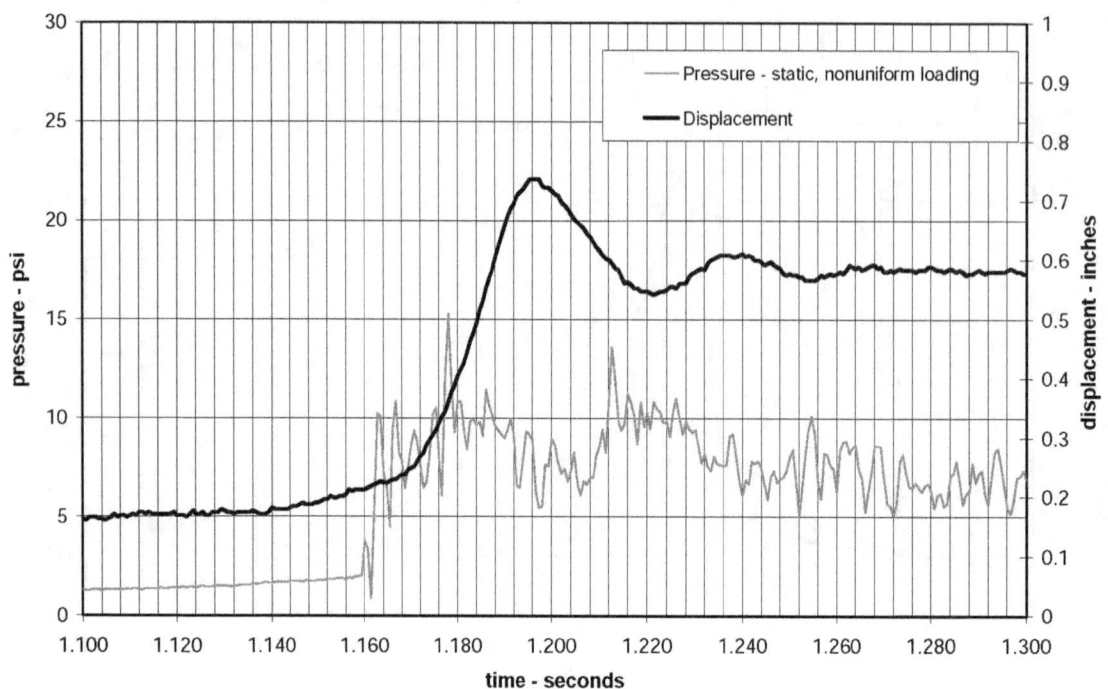

Figure A-93.—Category 3C - structure #8 - test 8 - static, nonuniform loading. Solid-concrete-block ventilation stopping - LLEM test #517.

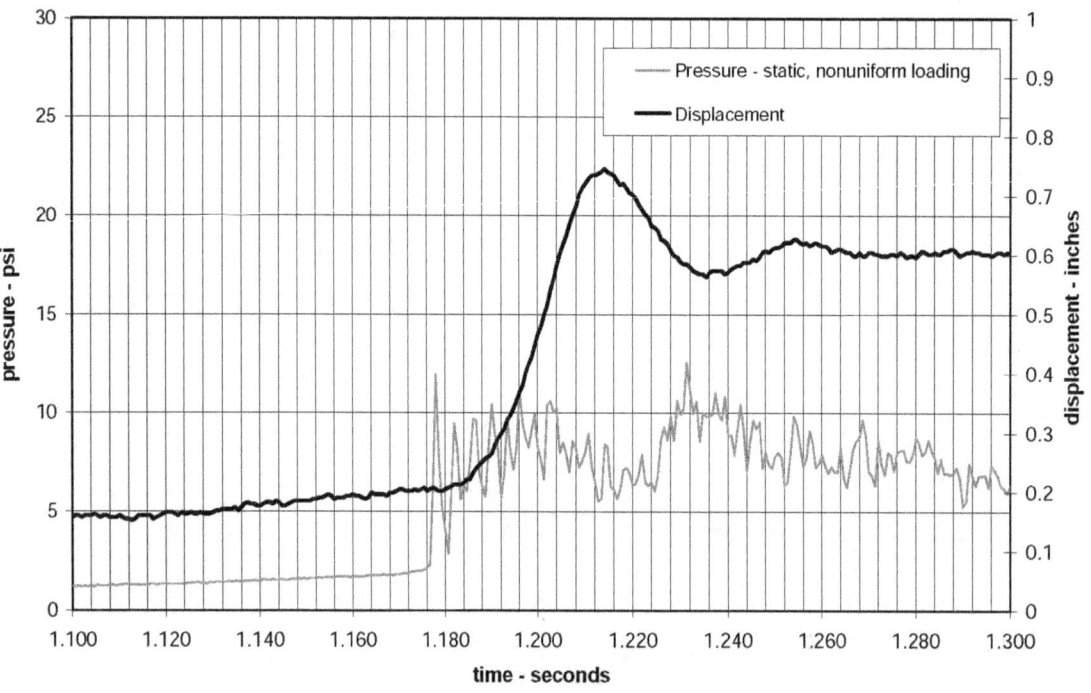

Figure A-94.—Category 3C - structure #8 - test 9 - static, nonuniform loading. Solid-concrete-block ventilation stopping - LLEM test #518.

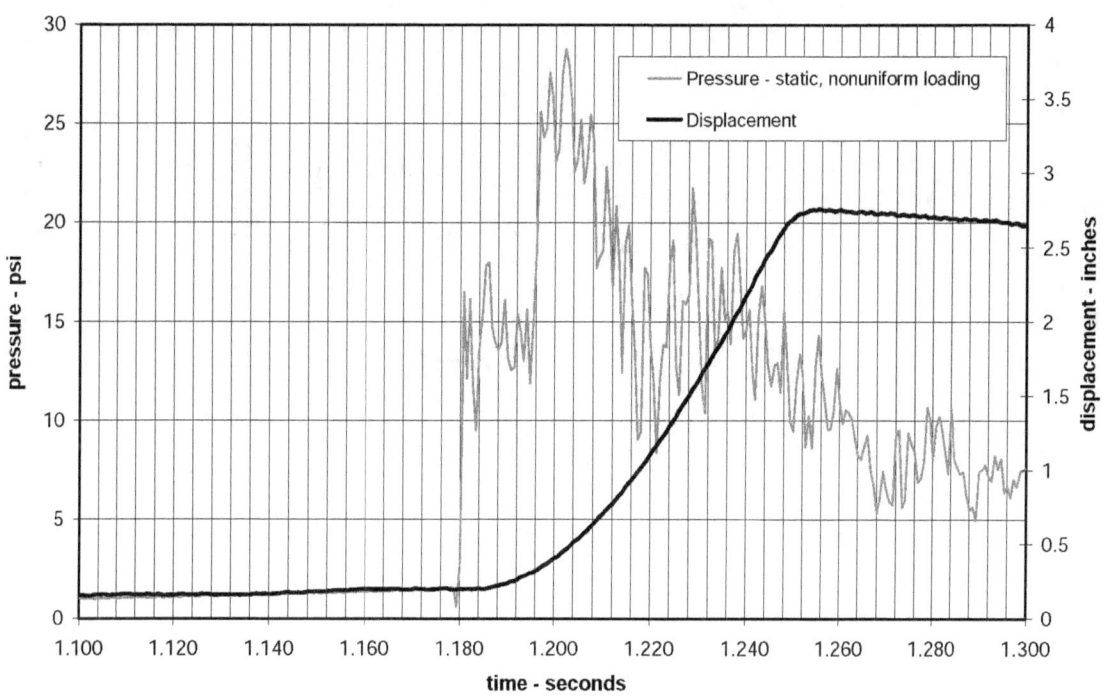

Figure A-95.—Category 3C - structure #8 - test 10 - static, nonuniform loading. Solid-concrete-block ventilation stopping - LLEM test #519.

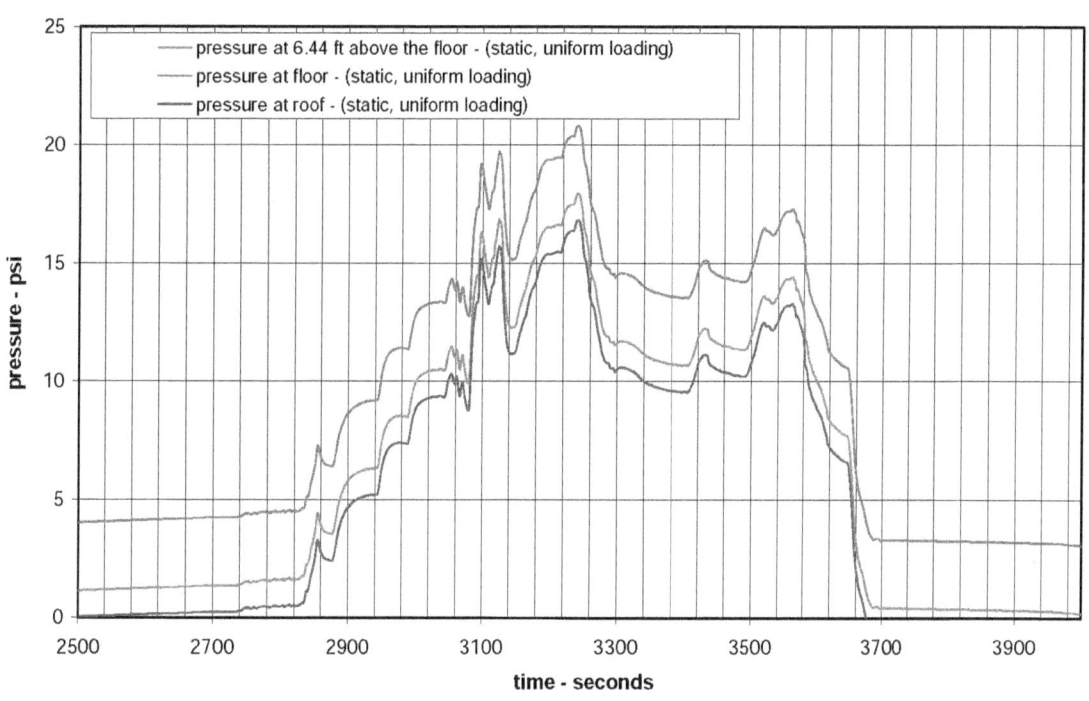

Figure A-96.—Category 4 - structure #1 - test 1 - static, uniform loading. Polymer and aggregate seal - test C8.

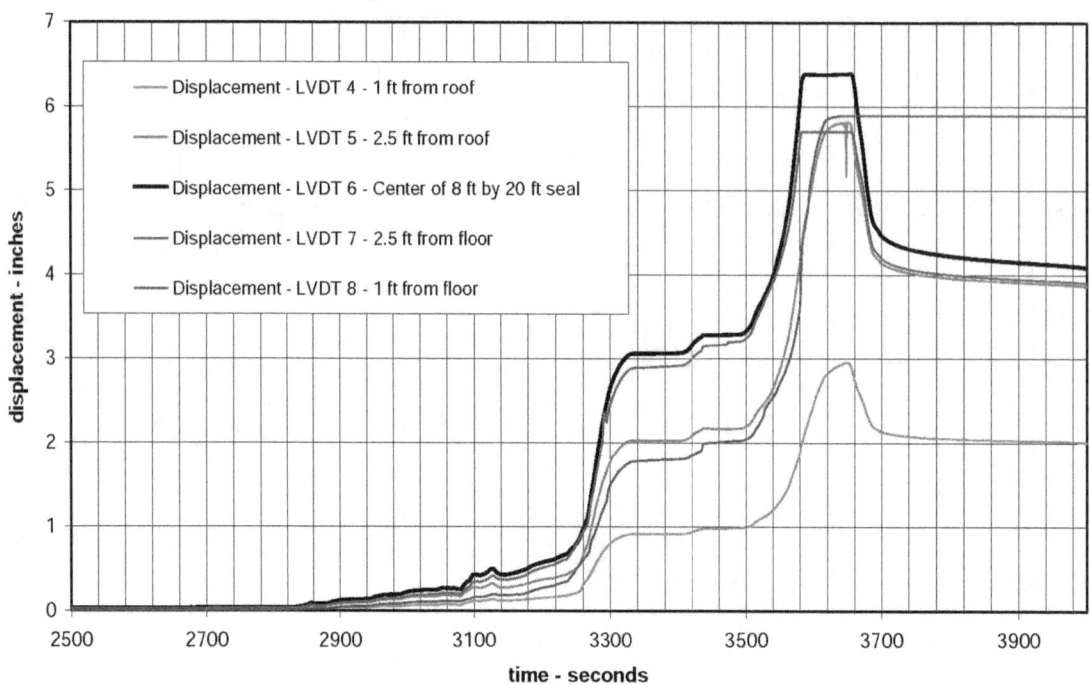

Figure A-97.—Category 4 - structure #1 - test 1 - static, uniform loading. Polymer and aggregate seal - test C8.

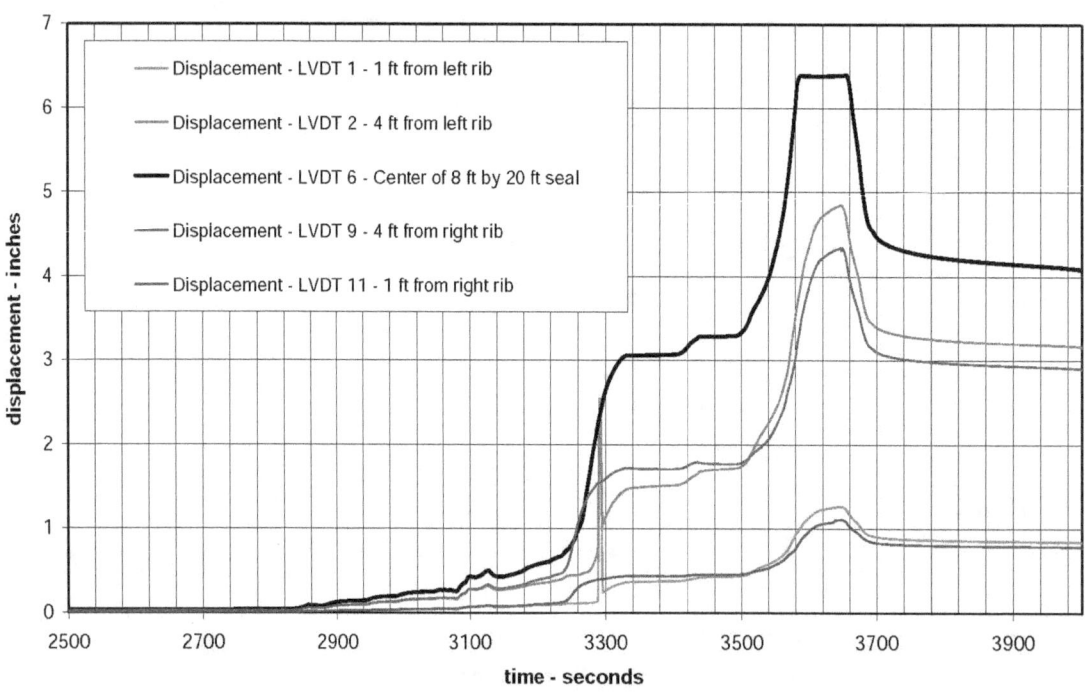

Figure A-98.—Category 4 - structure #1 - test 1 - static, uniform loading. Polymer and aggregate seal - test C8.

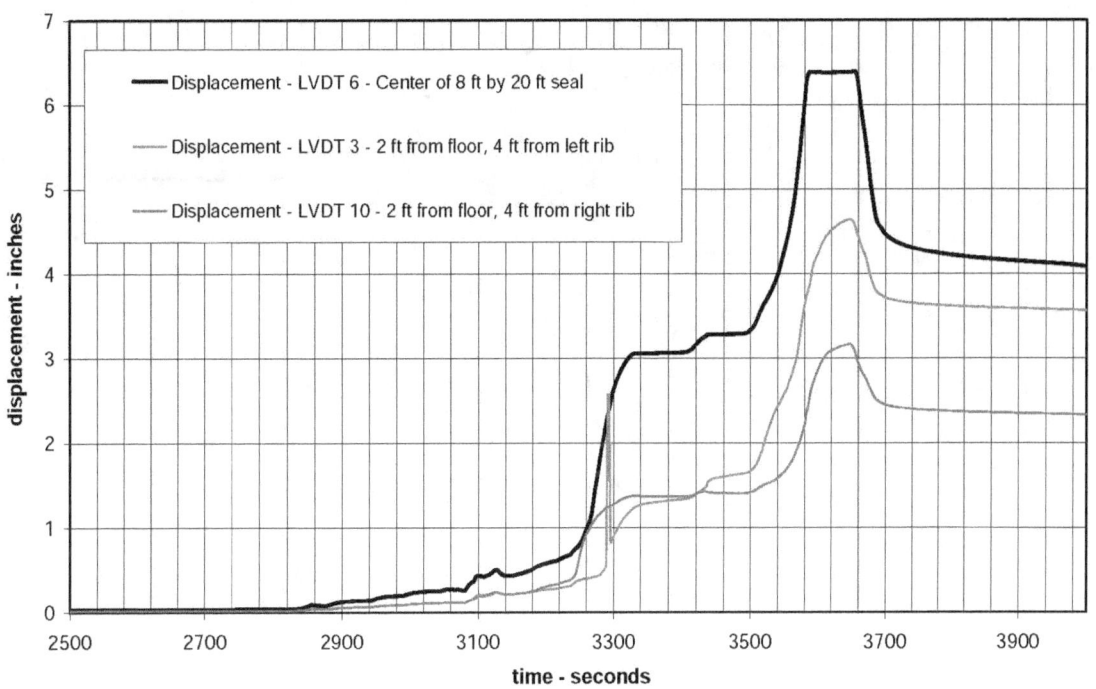

Figure A-99.—Category 4 - structure #1 - test 1 - static, uniform loading. Polymer and aggregate seal - test C8.

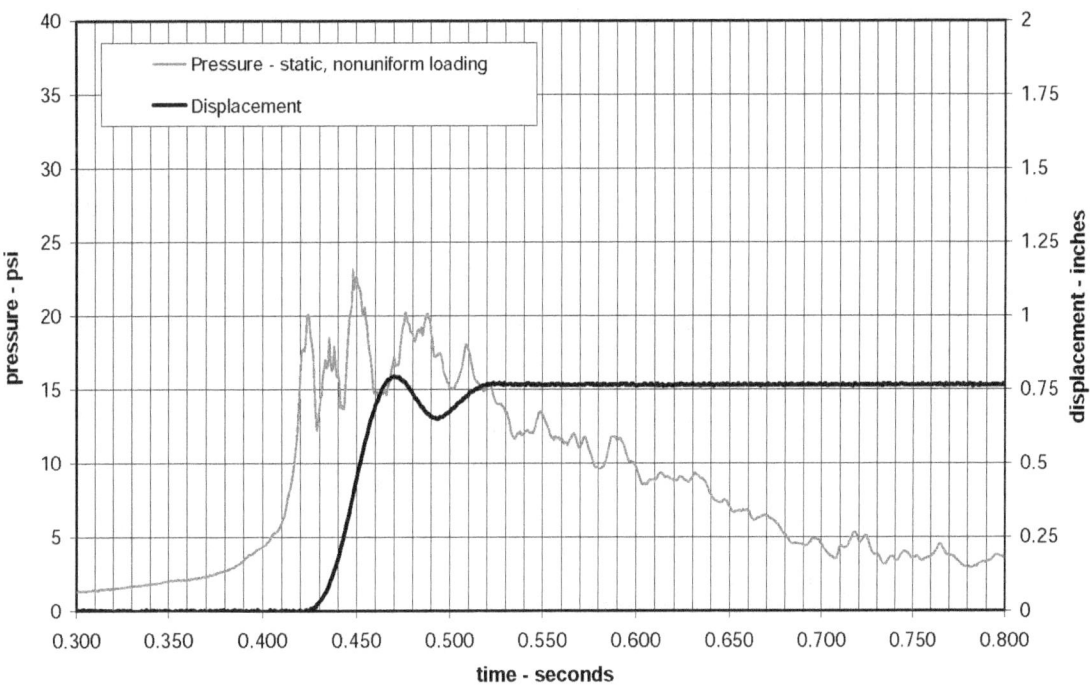

Figure A-100.—Category 5B - structure #1 - test 1 - static, nonuniform loading. Wood-crib-block seal with Packsetter bags - LLEM test #396.

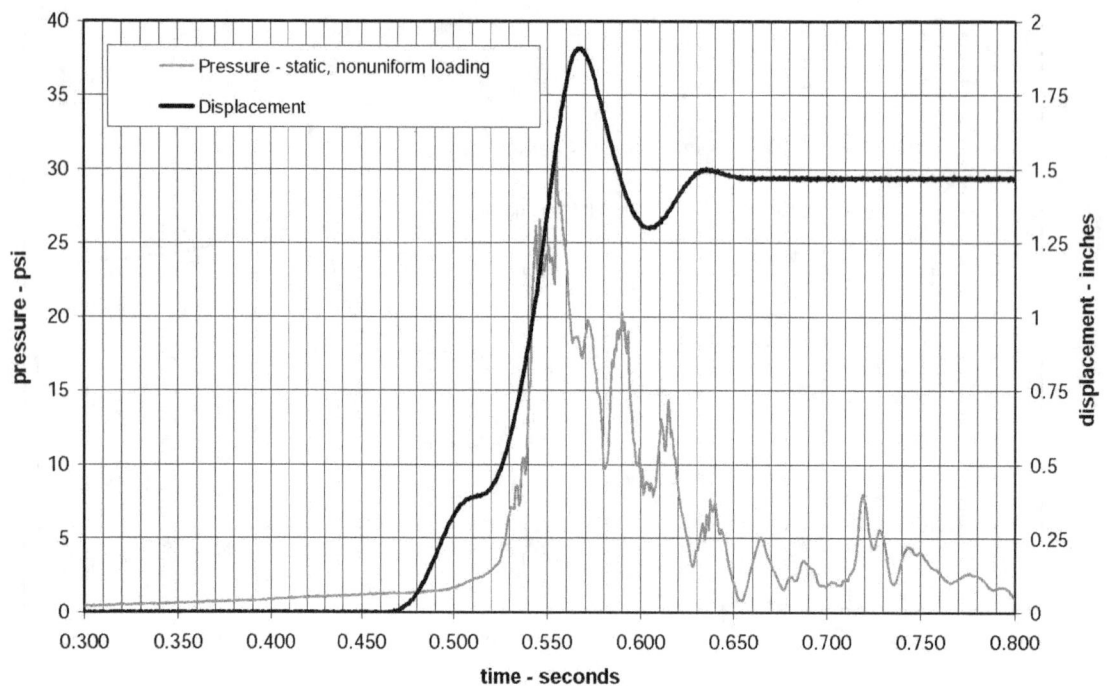

Figure A-101.—Category 5B - structure #1 - test 2 - static, nonuniform loading. Wood-crib-block seal with Packsetter bags - LLEM test #399.

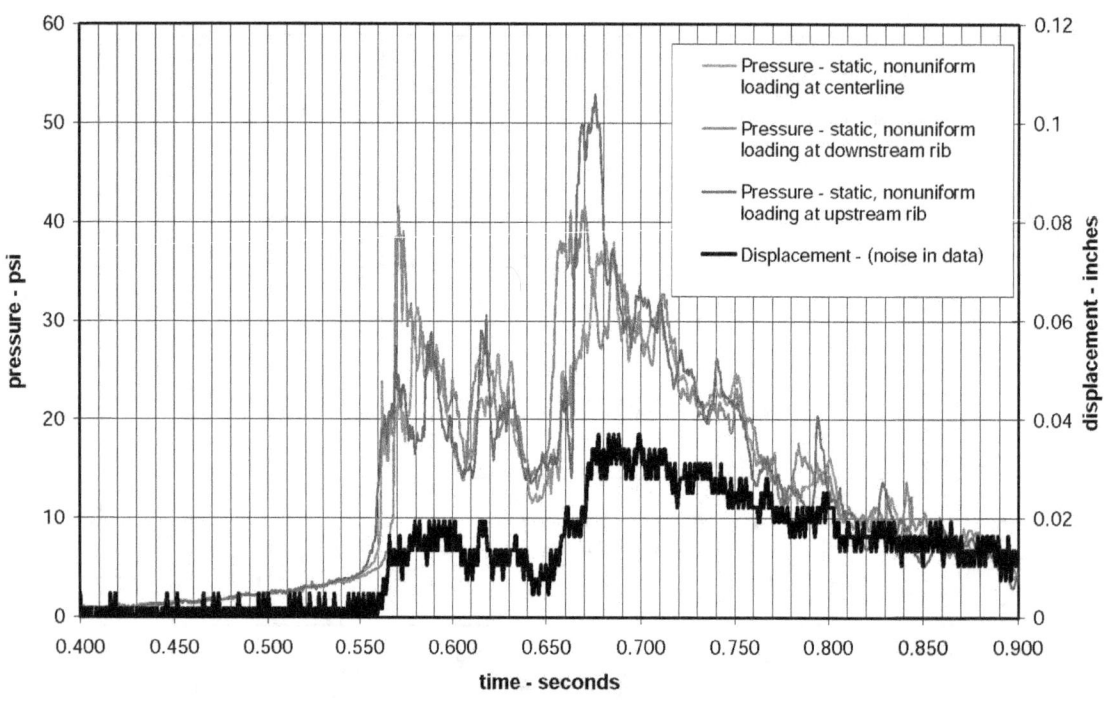

Figure A-102.—Category 6A - structure #3 - test 1 - static, nonuniform loading. Lightweight blocks - 24 in with hitching - LLEM test #508.

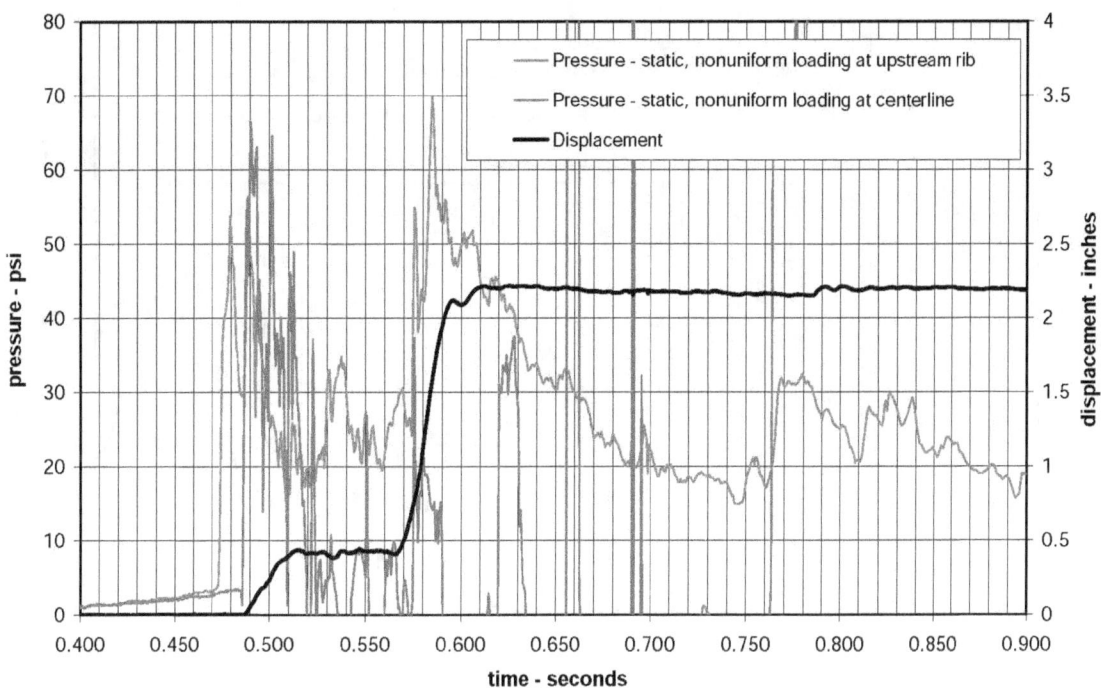

Figure A-103.—Category 6A - structure #3 - test 2 - static, nonuniform loading.
Lightweight blocks - 24 in with hitching - LLEM test #509.

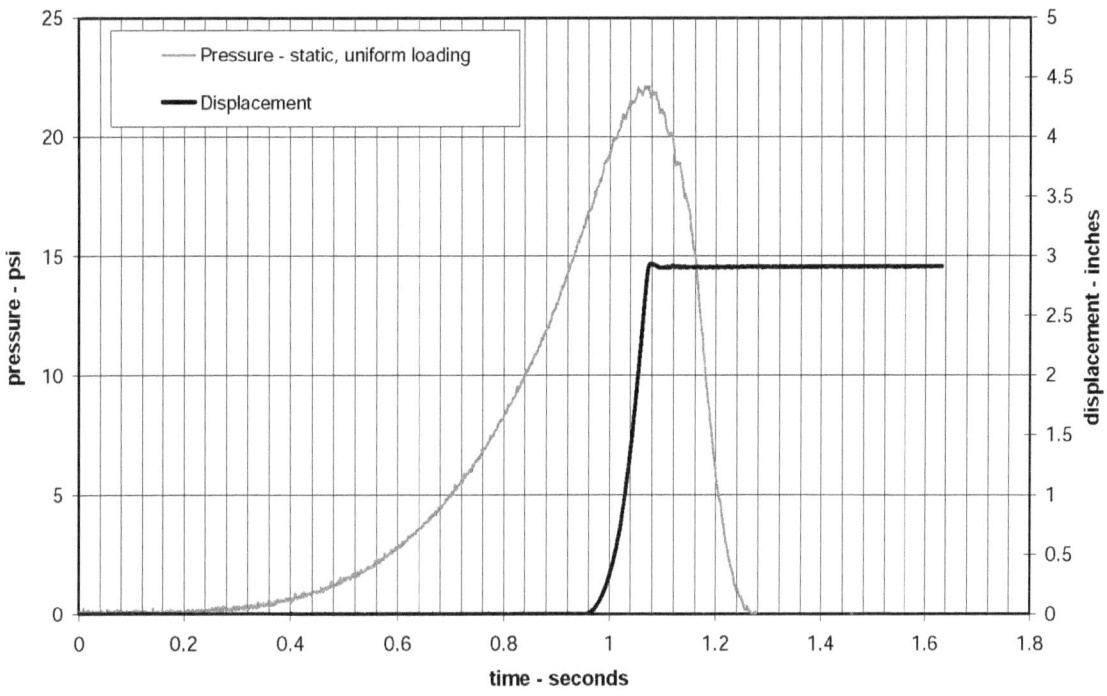

Figure A-104.—Category 6A - structure #2 - test 1 - static, uniform loading.
Lightweight blocks - 24 in with hitching - test 4-48.

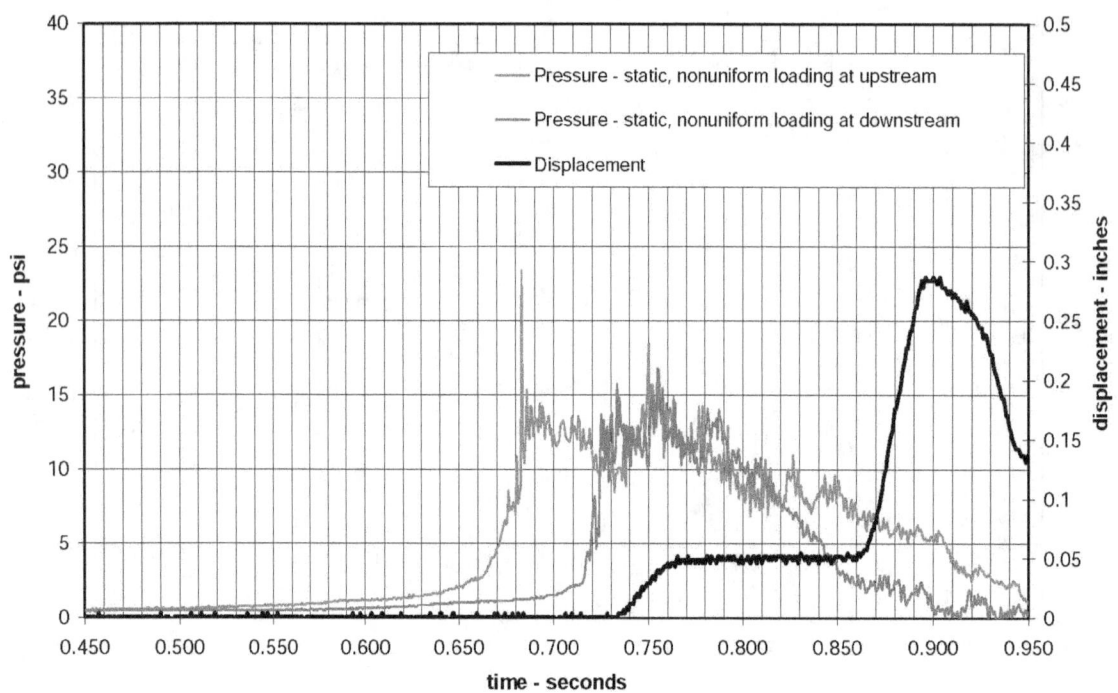

Figure A-105.—Category 6B - structure #1 - test 1 - static, nonuniform loading. Lightweight blocks - 40 in, no hitching - LLEM test #403.

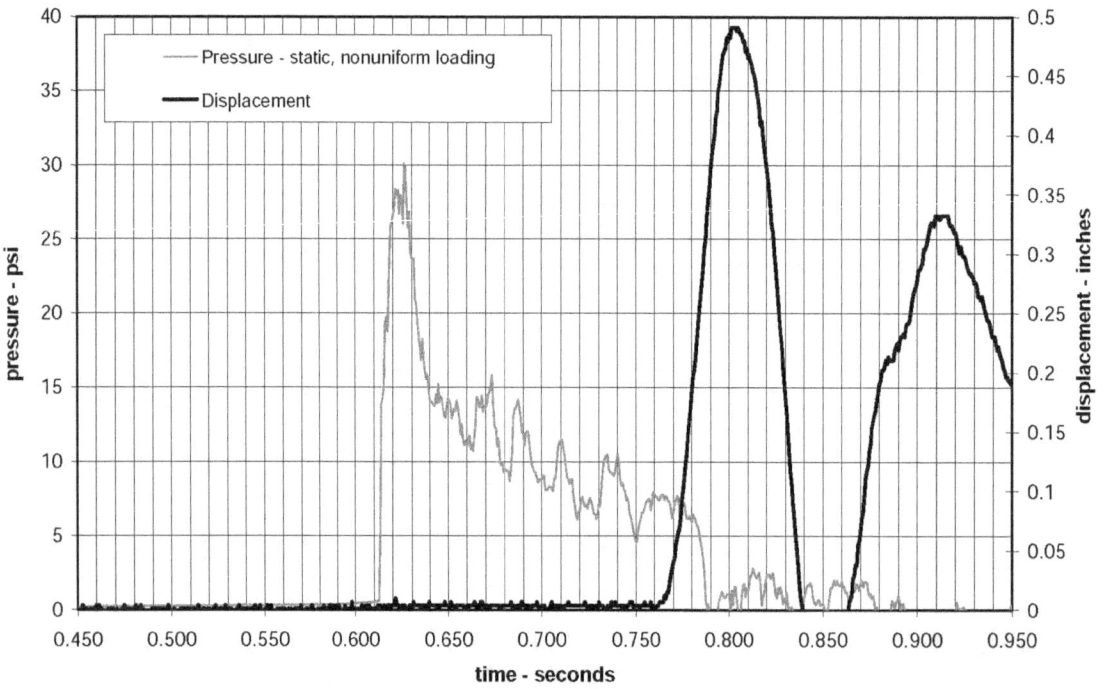

Figure A-106.—Category 6B - structure #1 - test 2 - static, nonuniform loading. Lightweight blocks - 40 in, no hitching - LLEM test #404.

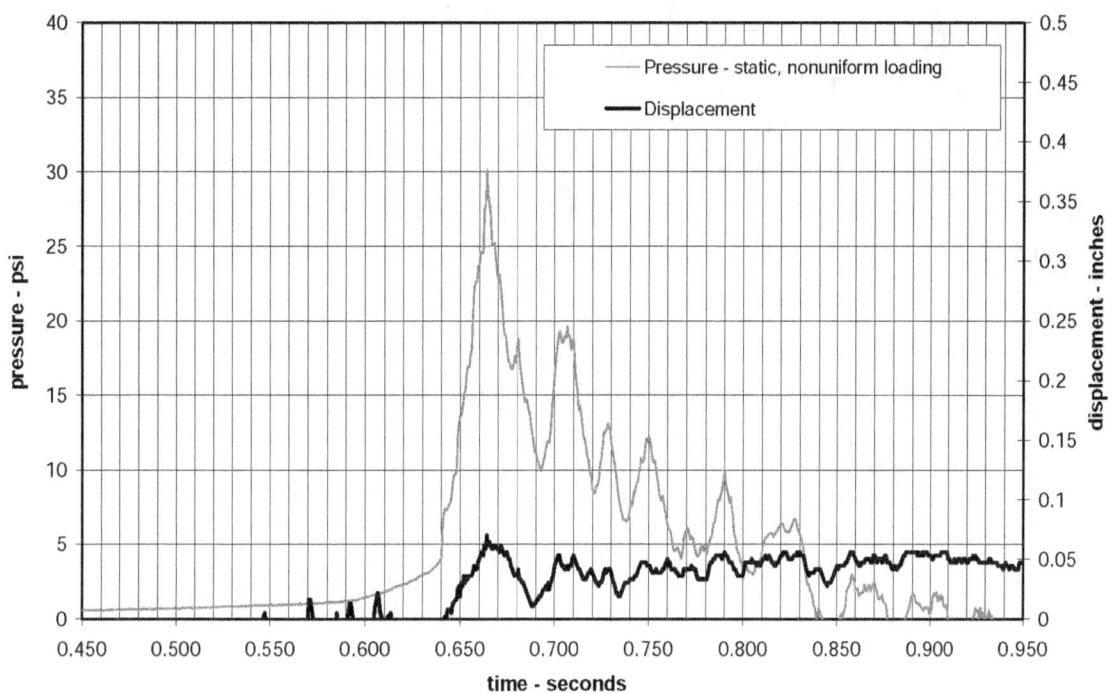

Figure A-107.—Category 6B - structure #1 - test 3 - static, nonuniform loading. Lightweight blocks - 40 in, no hitching - LLEM test #405.

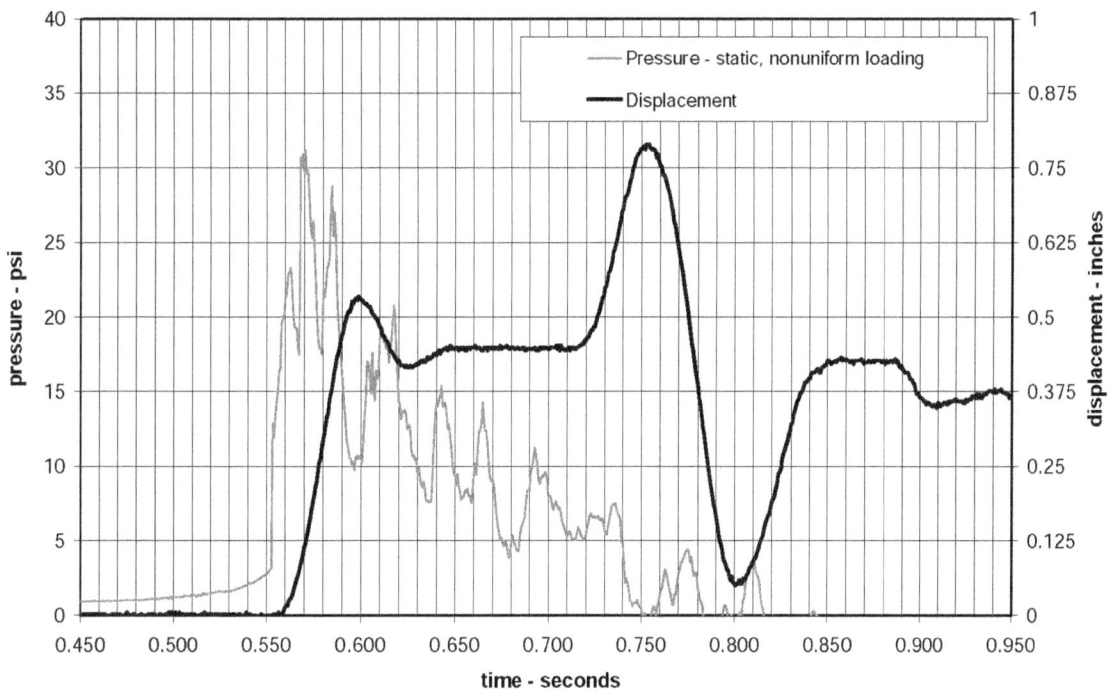

Figure A-108.—Category 6B - structure #1 - test 4 - static, nonuniform loading. Lightweight blocks - 40 in, no hitching - LLEM test #406.

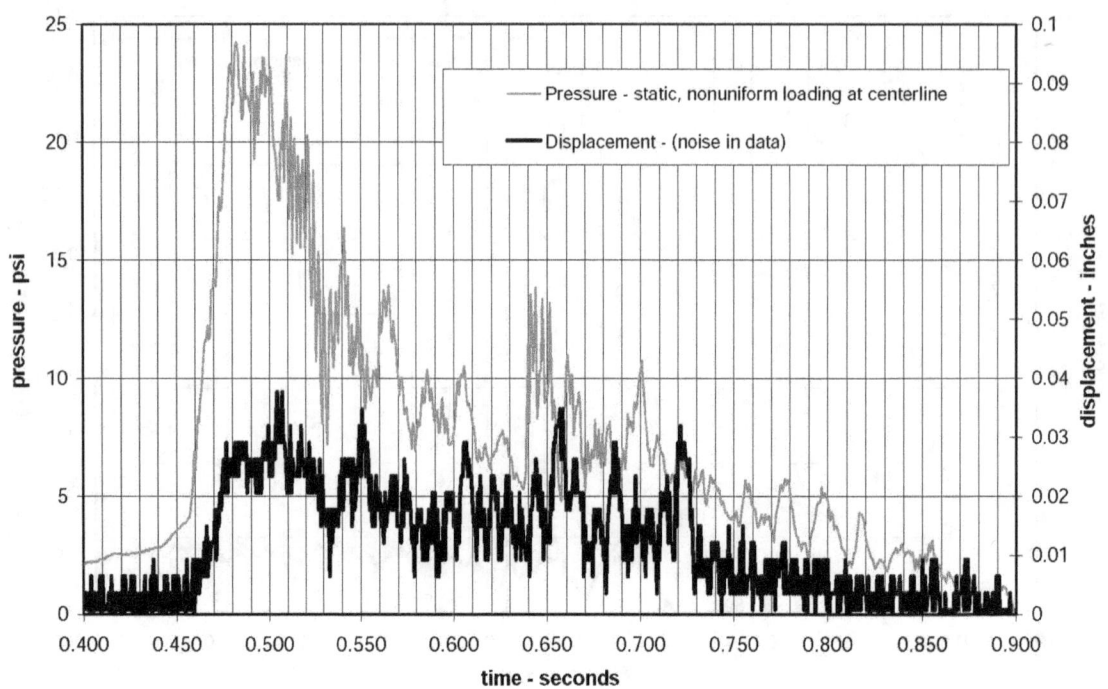

Figure A-109.—Category 6B - structure #2 - test 1 - static, nonuniform loading. Lightweight blocks - 40 in, no hitching - LLEM test #501.

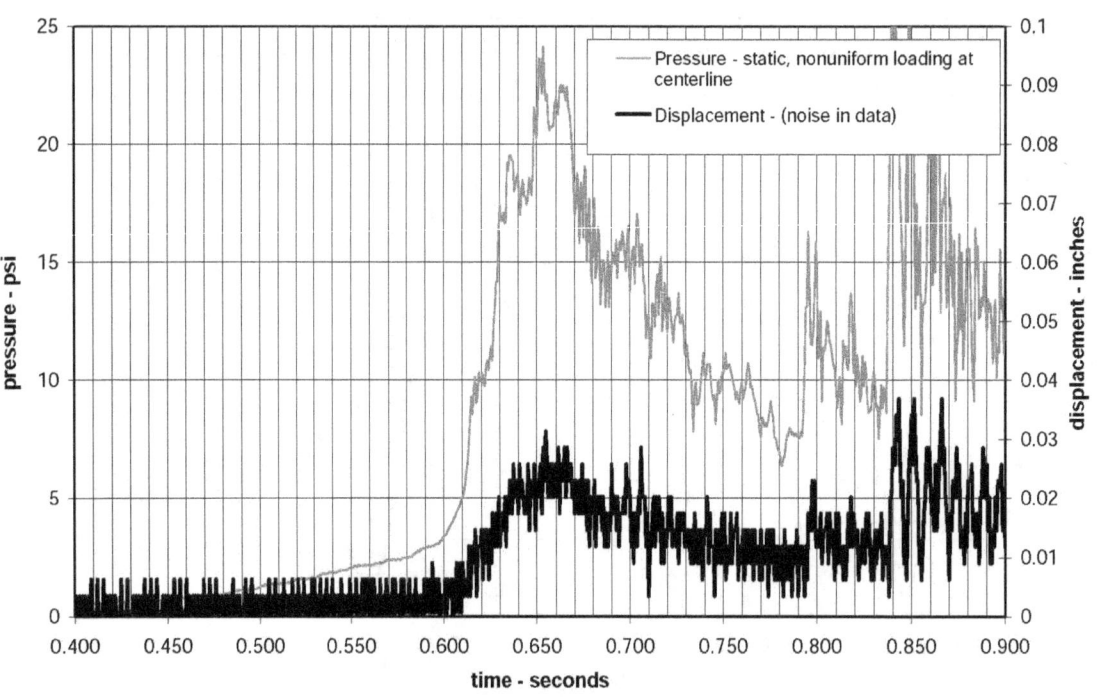

Figure A-110.—Category 6B - structure #2 - test 2 - static, nonuniform loading. Lightweight blocks - 40 in, no hitching - LLEM test #502.

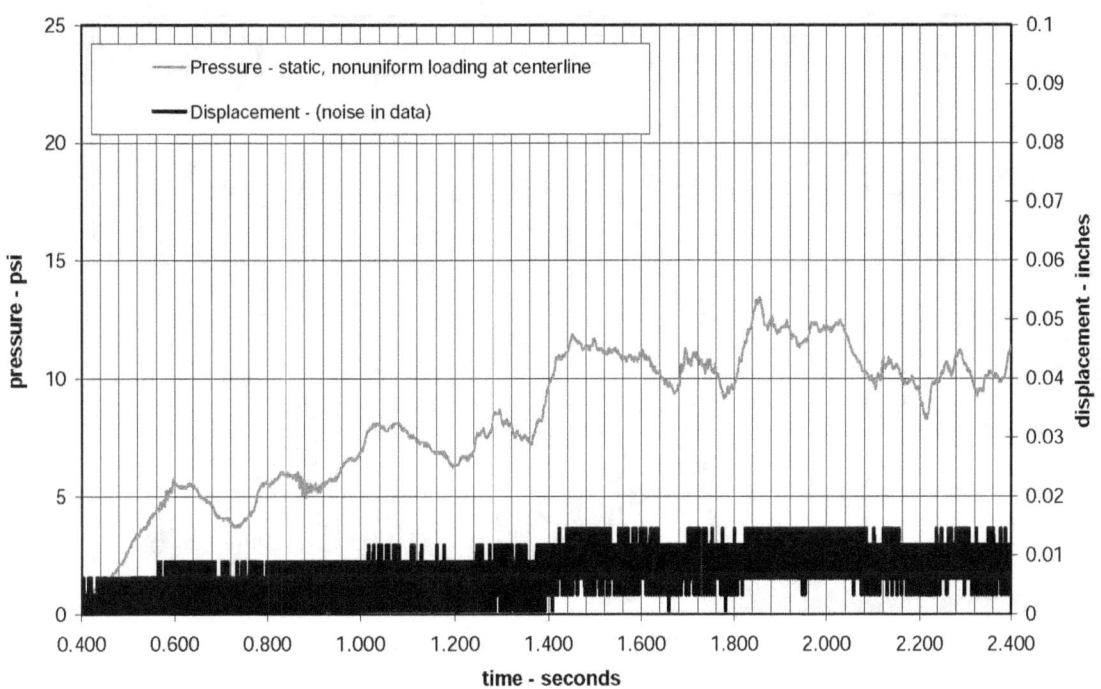

Figure A-111.—Category 6B - structure #2 - test 3 - static, nonuniform loading. Lightweight blocks - 40 in, no hitching - LLEM test #503.

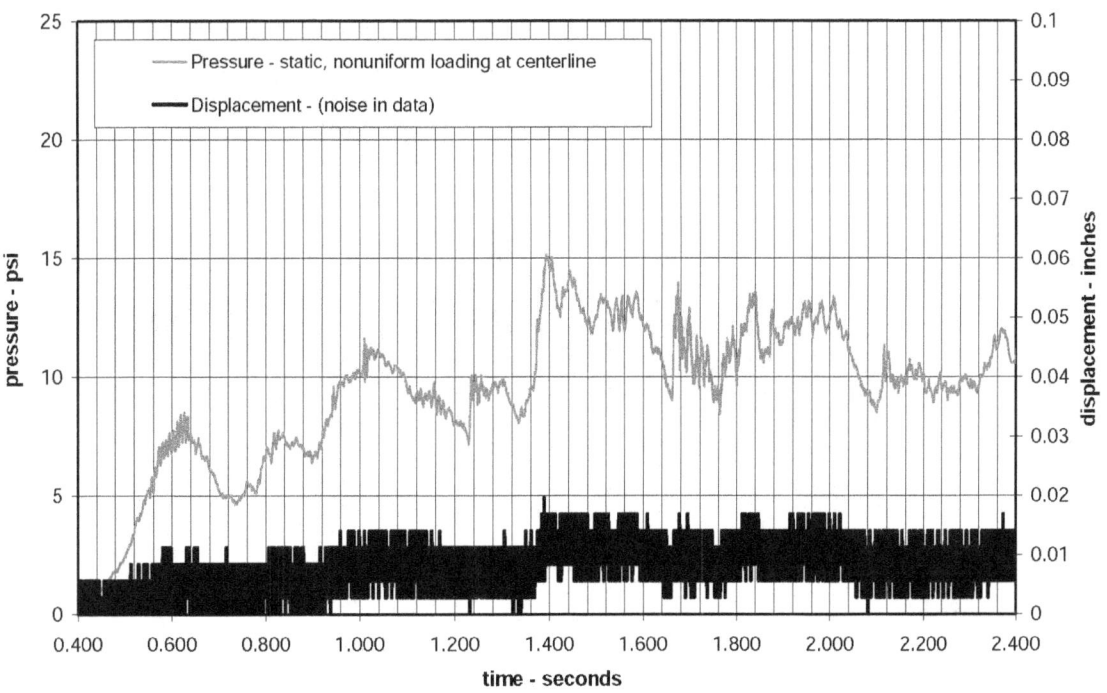

Figure A-112.—Category 6B - structure #2 - test 4 - static, nonuniform loading. Lightweight blocks - 40 in, no hitching - LLEM test #504.

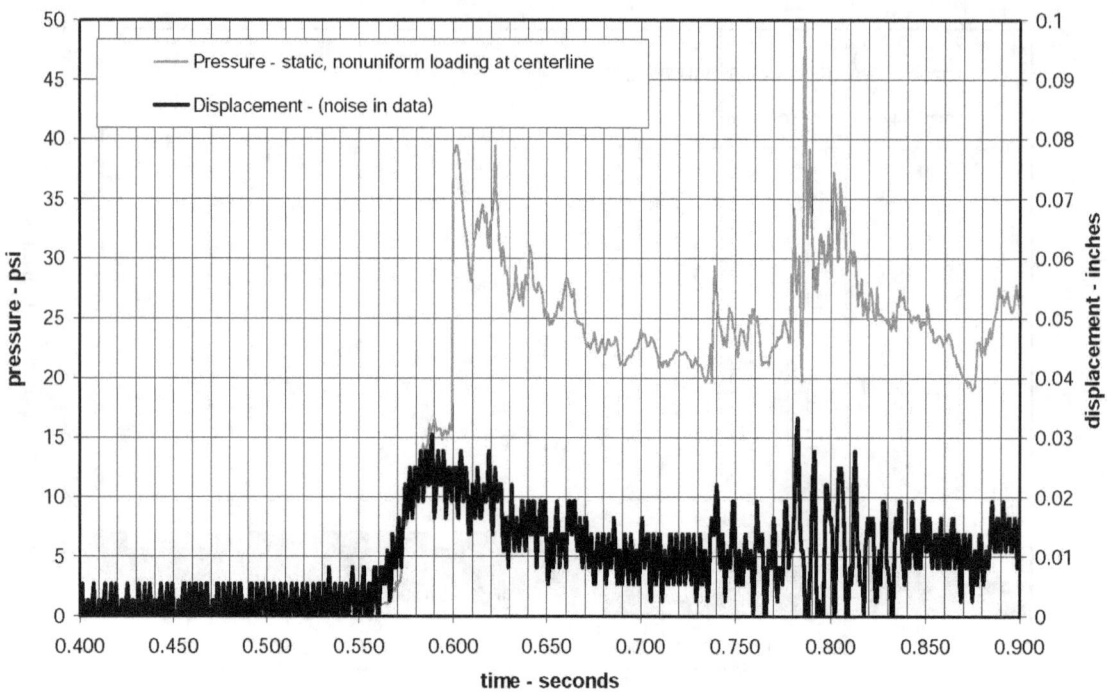

Figure A-113.—Category 6B - structure #2 - test 5 - static, nonuniform loading.
Lightweight blocks - 40 in, no hitching - LLEM test #505.

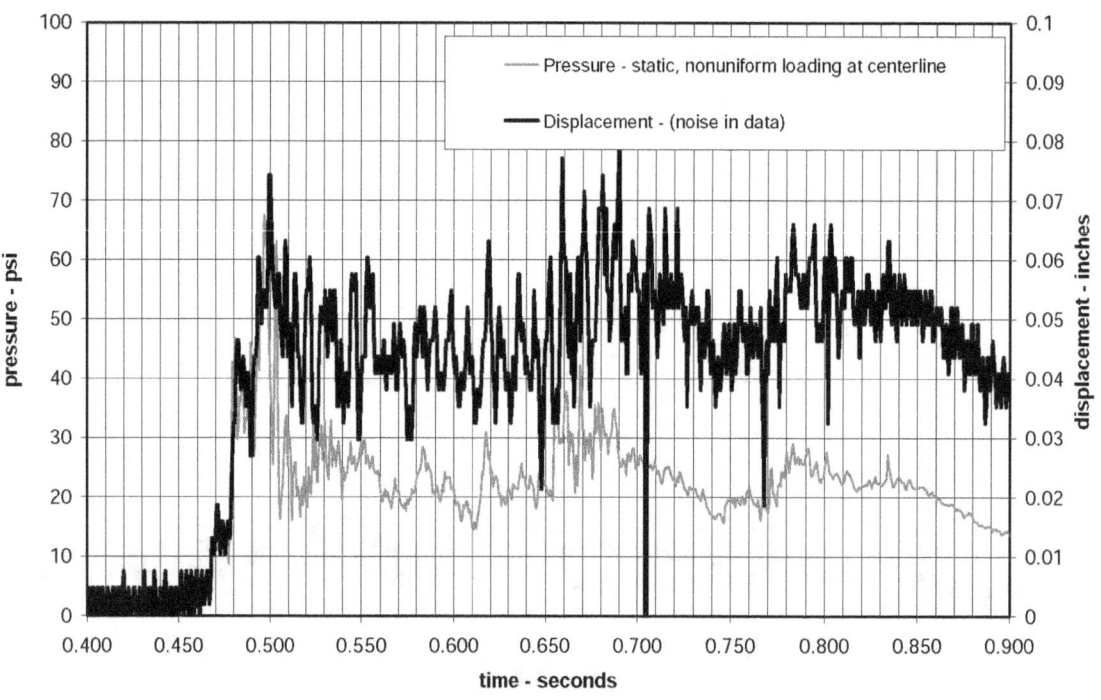

Figure A-114.—Category 6B - structure #2 - test 6 - static, nonuniform loading.
Lightweight blocks - 40 in, no hitching - LLEM test #506.

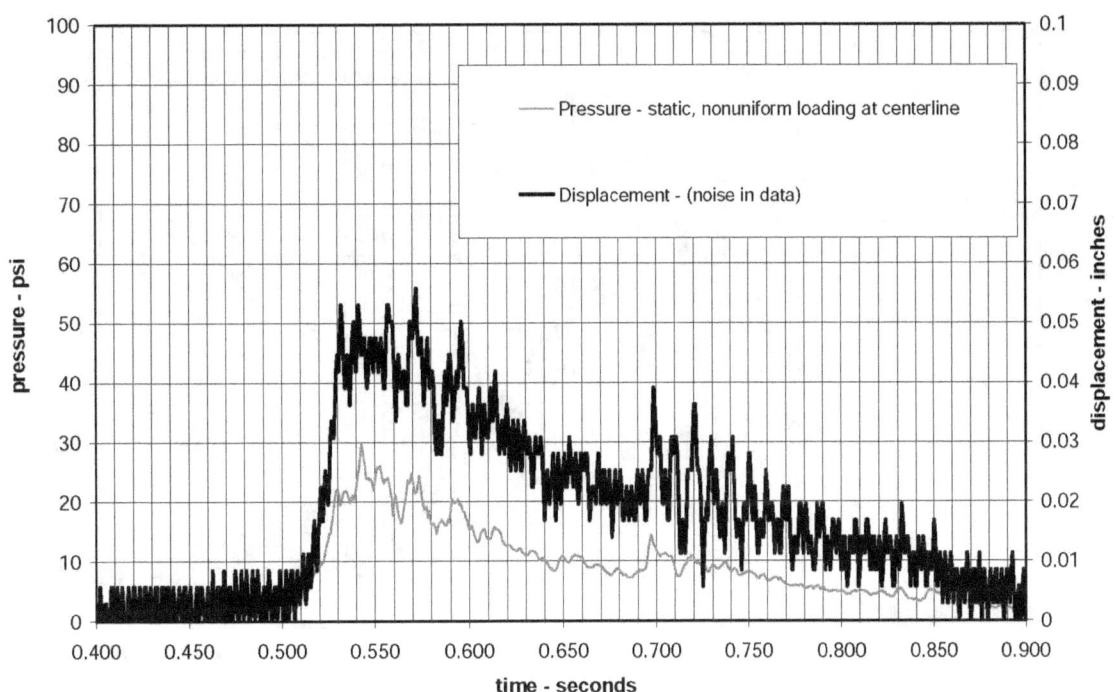

Figure A-115.—Category 6B - structure #2 - test 7 - static, nonuniform loading. Lightweight blocks - 40 in, no hitching - LLEM test #507.

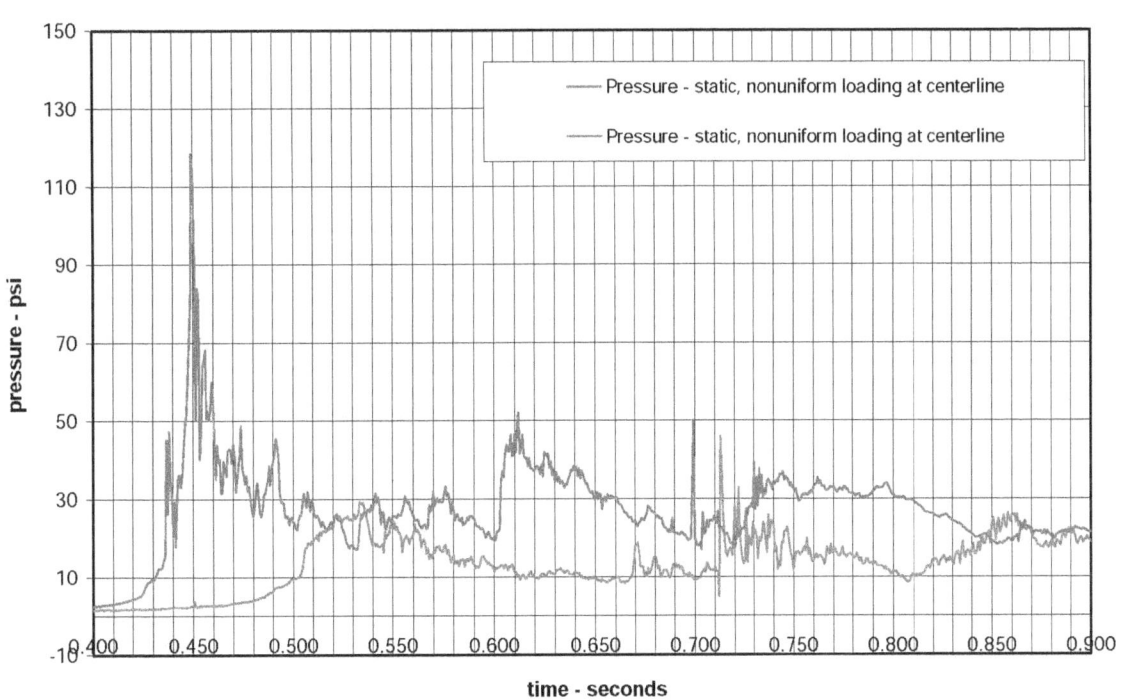

Figure A-116.—Category 6B - structure #2 - tests 8 and 9 - static, nonuniform loading. Lightweight blocks - 40 in, no hitching - LLEM tests #508–509.

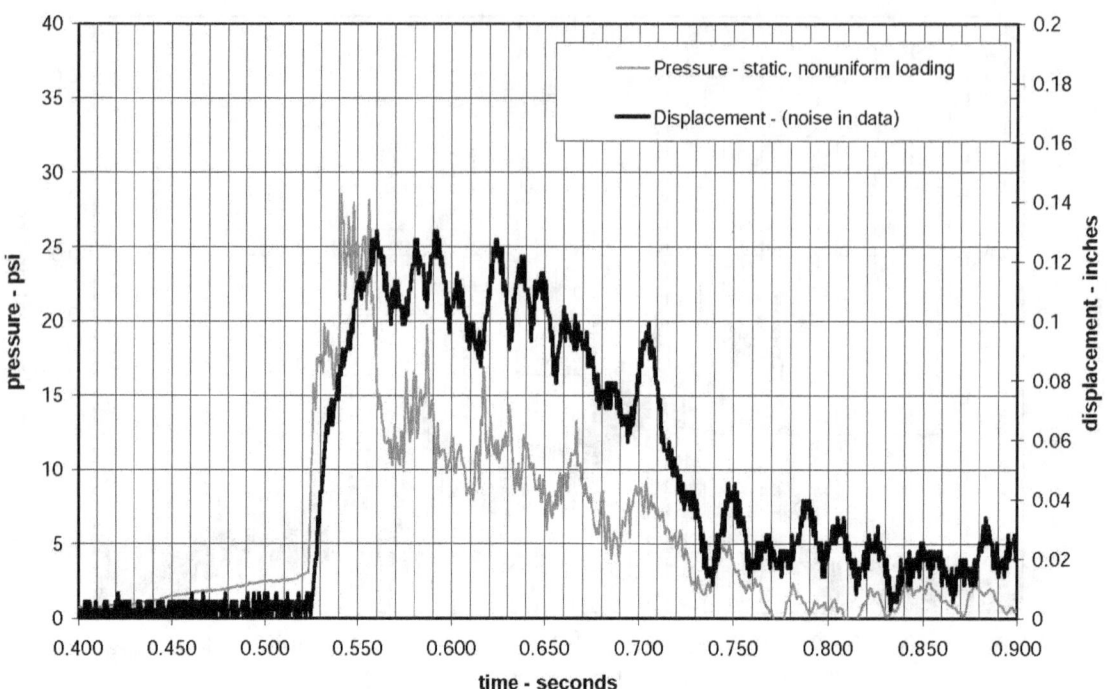

Figure A-117.—Category 6B - structure #3 - test 1 - static, nonuniform loading.
Lightweight blocks - 40 in, no hitching - LLEM test #501.

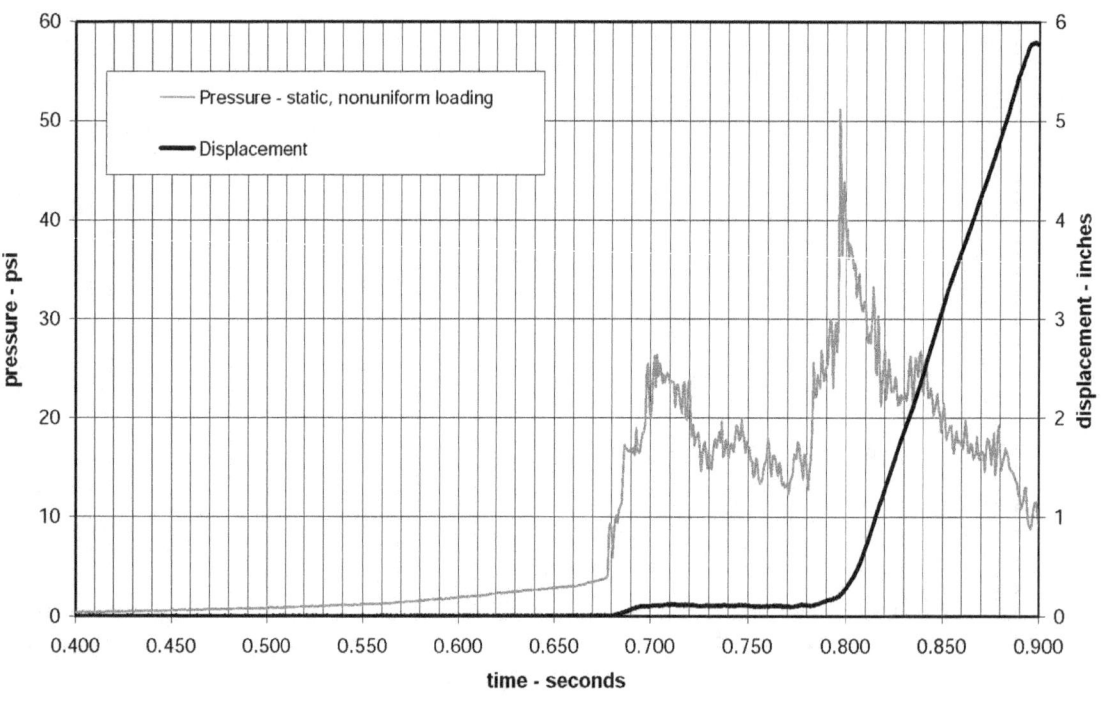

Figure A-118.—Category 6B - structure #3 - test 2 - static, nonuniform loading.
Lightweight blocks - 40 in, no hitching - LLEM test #502.

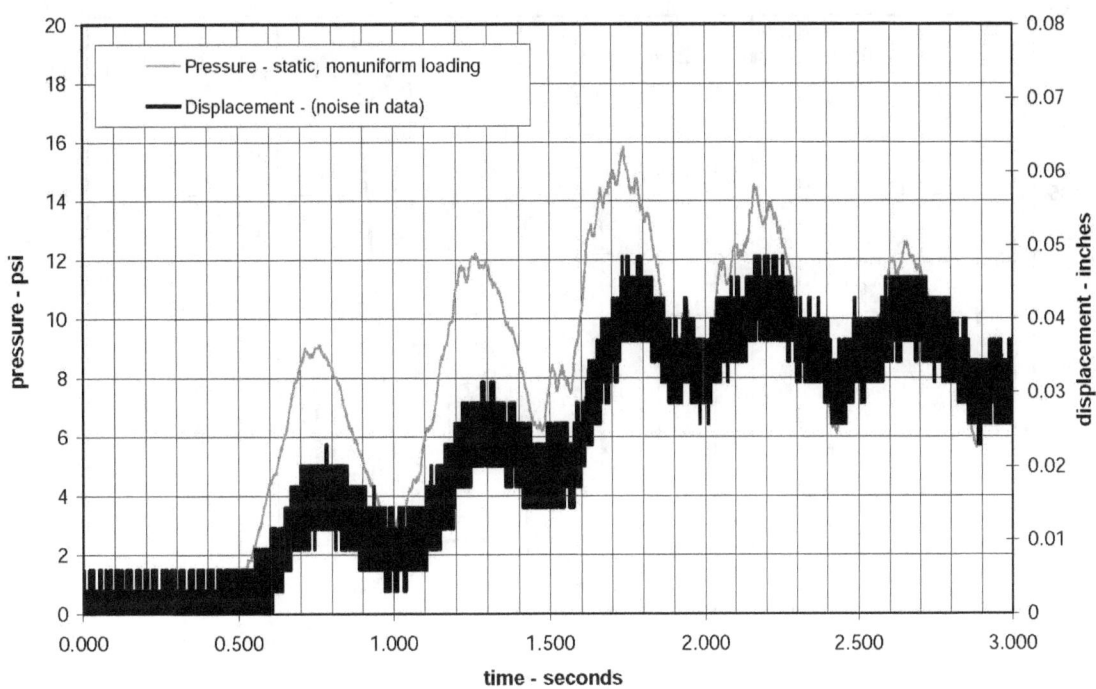

Figure A-119.—Category 6B - structure #4 - test 1 - static, nonuniform loading. Lightweight blocks - 40 in, no hitching - LLEM test #503.

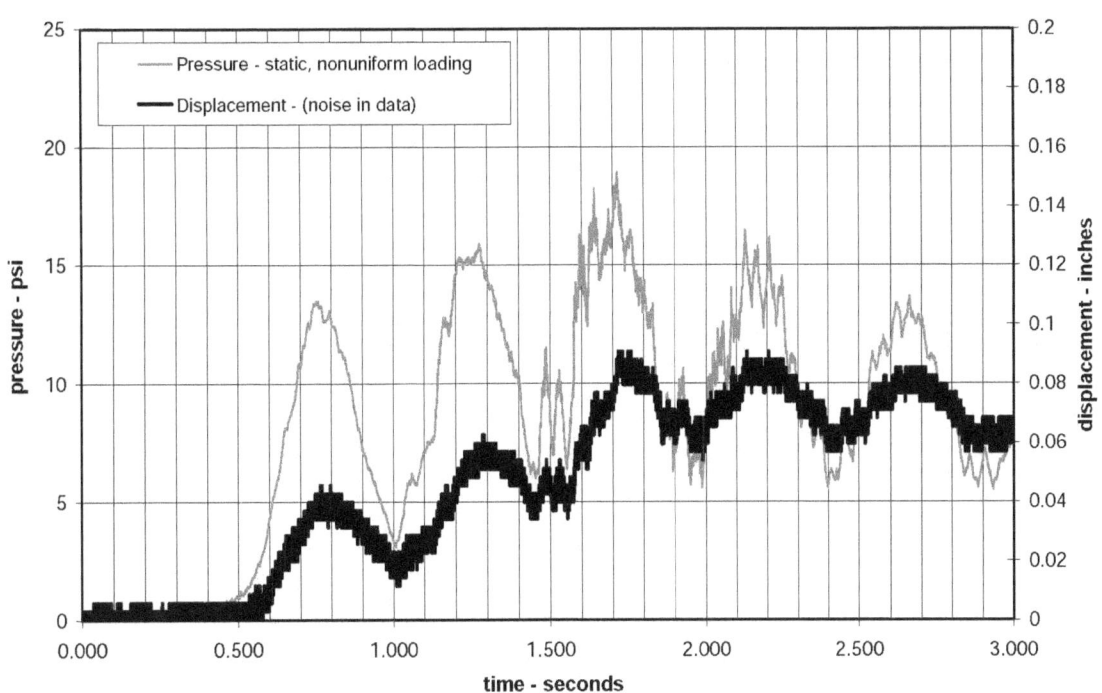

Figure A-120.—Category 6B - structure #4 - test 2 - static, nonuniform loading. Lightweight blocks - 40 in, no hitching - LLEM test #504.

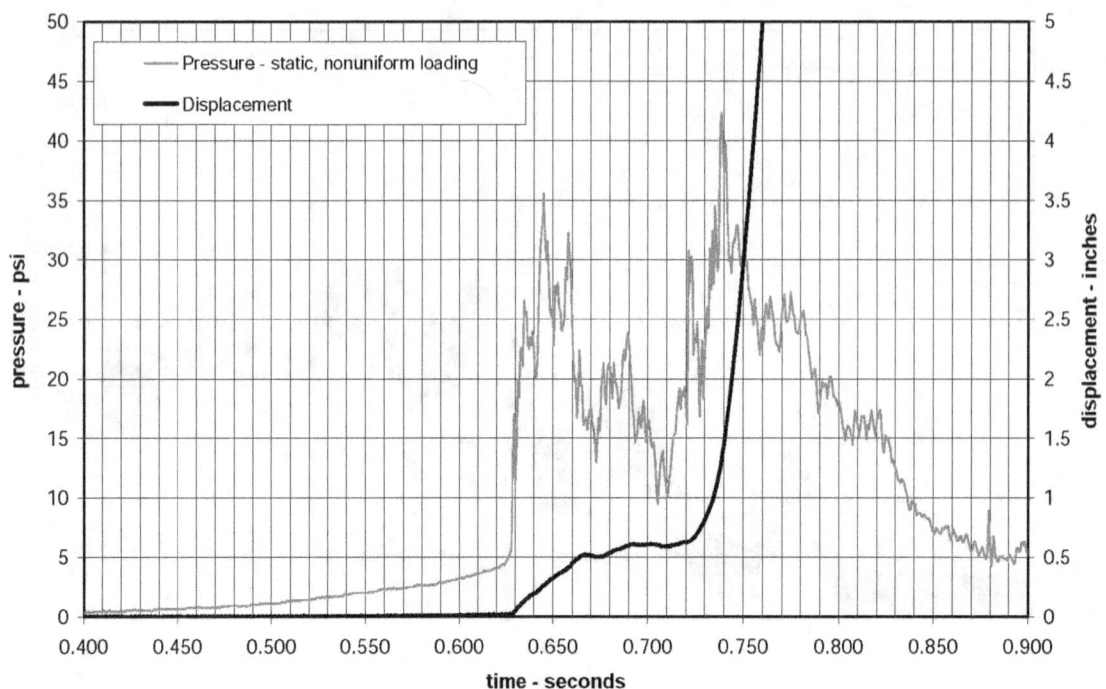

Figure A-121.—Category 6B - structure #4 - test 3 - static, nonuniform loading.
Lightweight blocks - 40 in, no hitching - LLEM test #505.

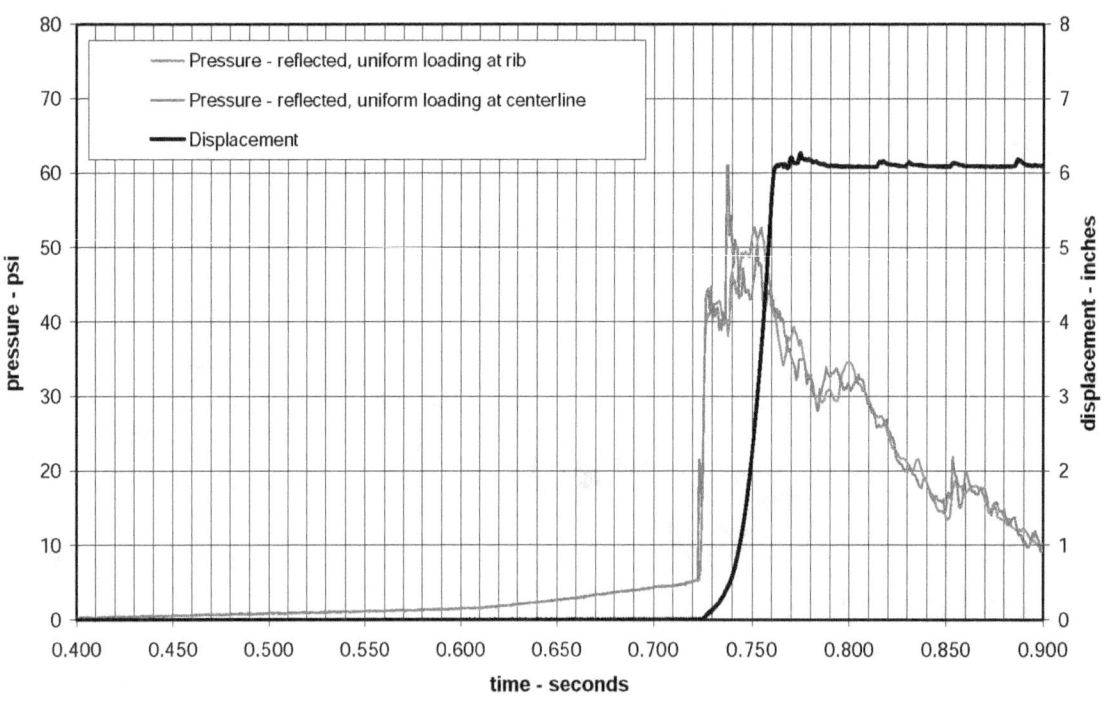

Figure A-122.—Category 6B - structure #5 - test 1 - reflected, uniform loading.
Lightweight blocks - 40 in, no hitching - LLEM test #502.

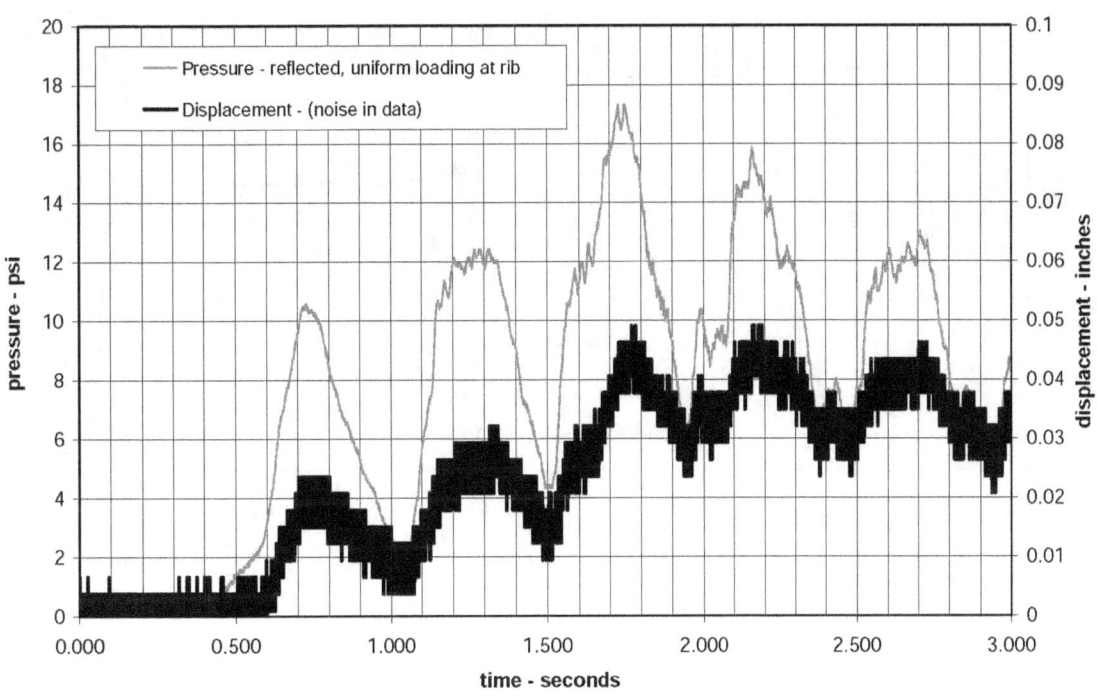

Figure A-123.—Category 6B - structure #6 - test 1 - reflected, uniform loading. Lightweight blocks - 40 in, no hitching - LLEM test #503.

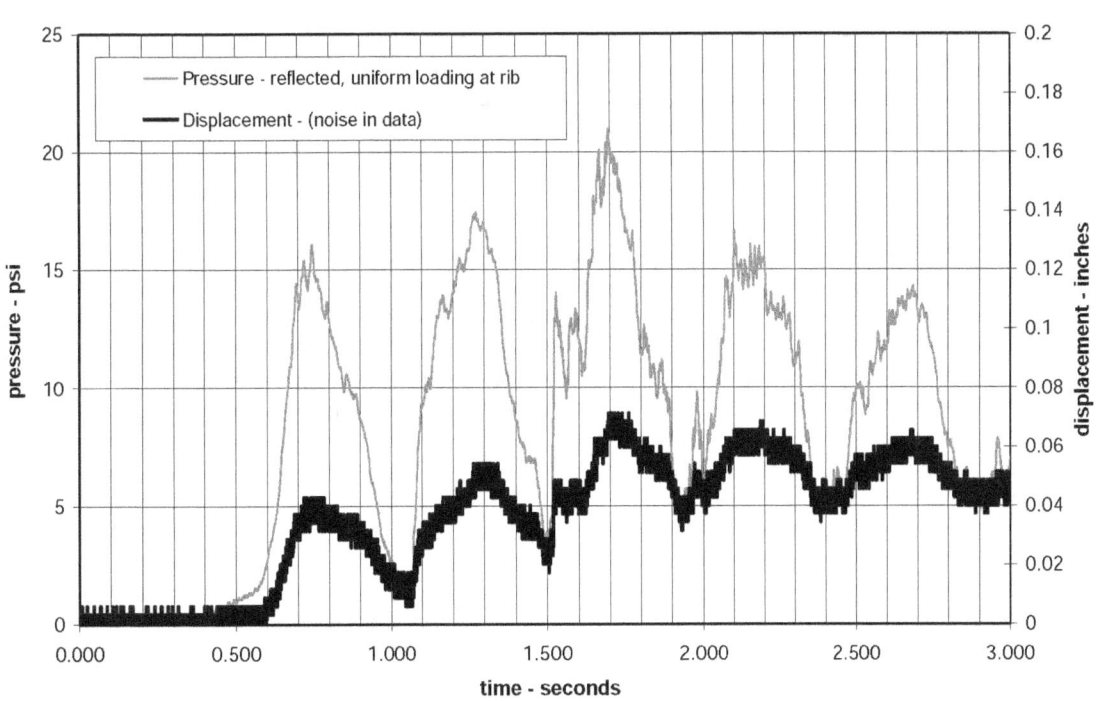

Figure A-124.—Category 6B - structure #6 - test 2 - reflected, uniform loading. Lightweight blocks - 40 in, no hitching - LLEM test #504.

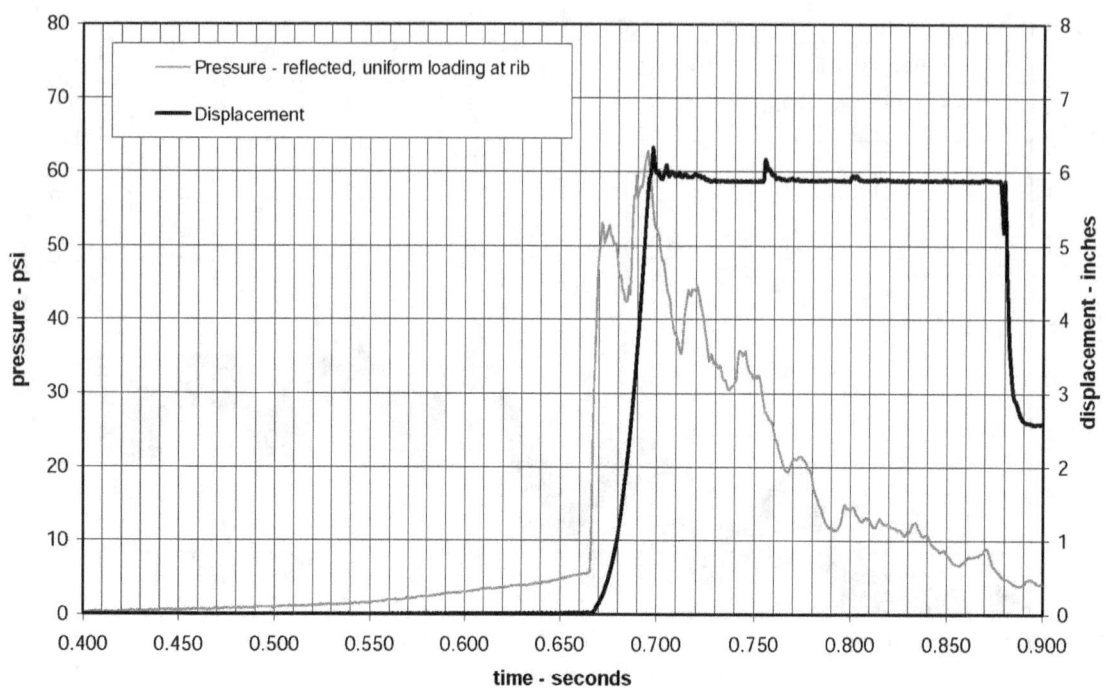

Figure A-125.—Category 6B - structure #6 - test 3 - reflected, uniform loading. Lightweight blocks - 40 in, no hitching - LLEM test #505.

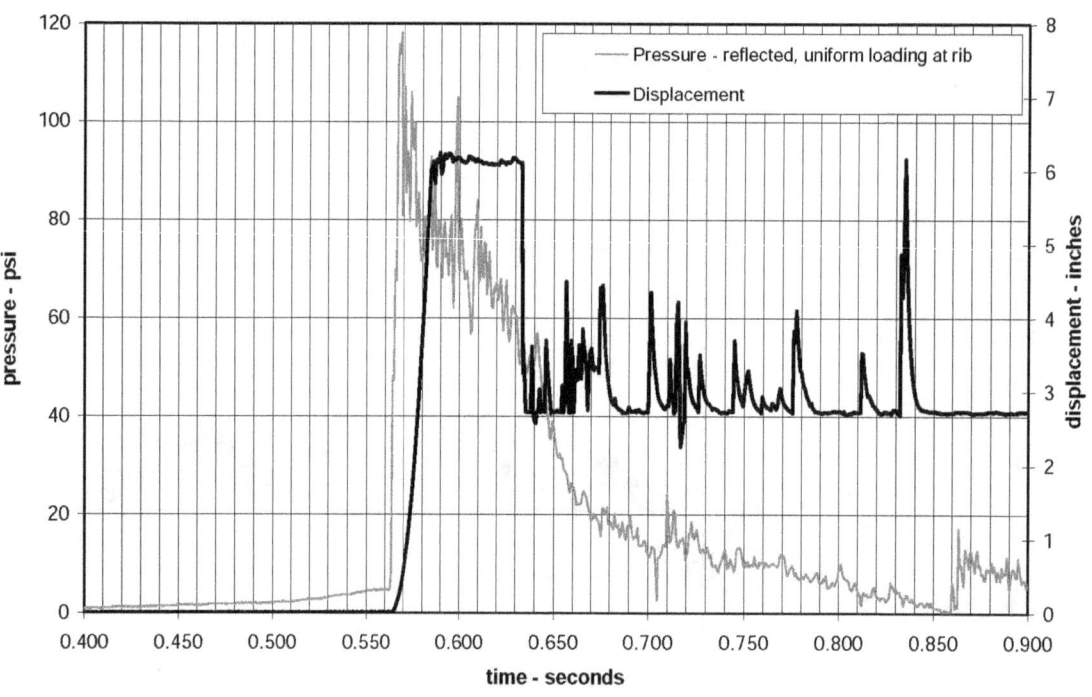

Figure A-126.—Category 6B - structure #7 - test 1 - reflected, uniform loading. Lightweight blocks - 40 in, no hitching - LLEM test #506.

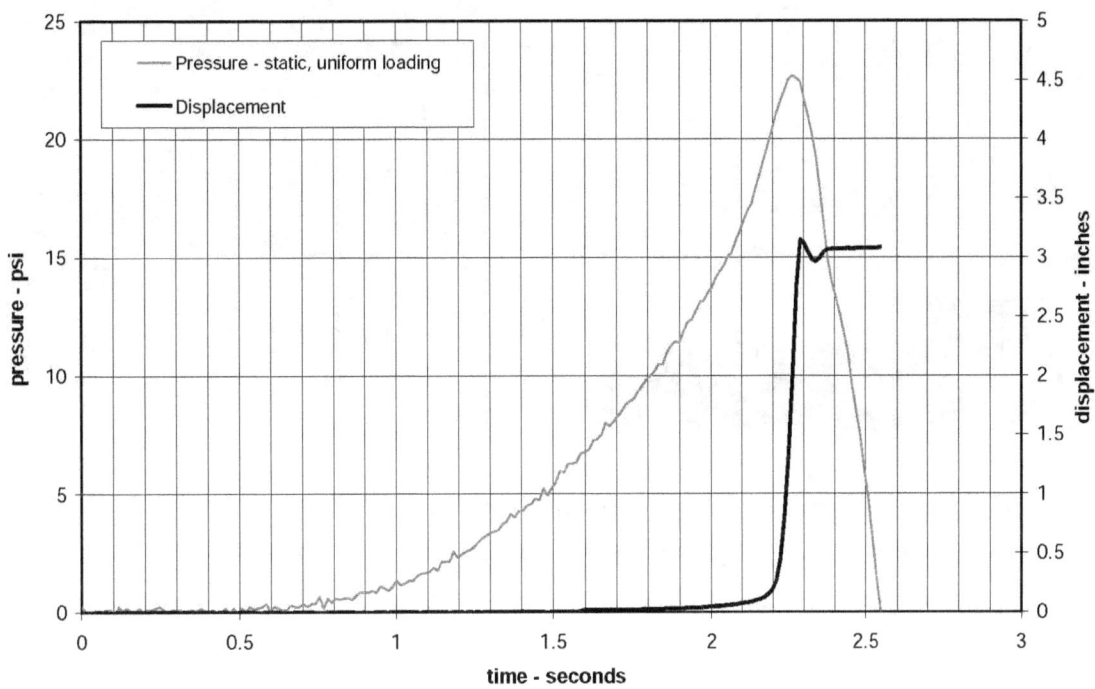

Figure A-127.—Category 6B - structure #8 - test 1 - static, uniform loading. Lightweight blocks - 40 in, no hitching - test C5-53E.

***Delivering on the Nation's promise:
safety and health at work for all people
through research and prevention***

To receive NIOSH documents or more information about occupational safety and health topics, contact NIOSH at

1–800–CDC–INFO (1–800–232–4636)
TTY: 1–888–232–6348
e-mail: cdcinfo@cdc.gov

or visit the NIOSH Web site at **www.cdc.gov/niosh.**

For a monthly update on news at NIOSH, subscribe to NIOSH *eNews* by visiting **www.cdc.gov/niosh/eNews.**

DHHS (NIOSH) Publication No. 2009-151

SAFER • HEALTHIER • PEOPLE™